D0851051

Academic

Preparation

In Mathematics

Teaching for Transition
From High School
To College

Library - St. Joseph's College
222 Clinton Avenue
Brooklyn, N.Y. 11205

College Entrance Examination Board, New York, 1985

Academic Preparation in Mathematics is one of a series of six books. The Academic Preparation Series includes books in English, the Arts, Mathematics, Science, Social Studies, and Foreign Language. Single copies of any one of these books can be purchased for $6.95. Orders for 5 to 49 copies receive a 20% discount; orders for 50 or more receive a 50% discount.

A boxed set of all the books in the Academic Preparation Series is available for $20.00. Orders for five or more sets receive a 20% discount. Each set also includes a copy of *Academic Preparation for College: What Students Need to Know and Be Able to Do*.

Payment or purchase order for individual titles or the set should be addressed to: College Board Publications, Box 886, New York, New York 10101.

Copyright © 1985 by College Entrance Examination Board. All rights reserved. The College Board and the acorn logo are registered trademarks of the College Entrance Examination Board. Educational EQuality Project is a trademark owned by the College Entrance Examination Board. Printed in the United States of America.

ISBN: 0-87447-222-9

9 8 7 6 5 4 3

117592

Contents

Principal Writer and Consultant

Jeremy Kilpatrick, Professor of Mathematics Education, University of Georgia

Mathematical Sciences Advisory Committee, 1984-85

John W. Kenelly, Clemson University, South Carolina (*Chair*)
Floyd L. Downs, Hillsdale High School, San Mateo, California
Carole E. Greenes, Boston University, Massachusetts
Jeremy Kilpatrick, University of Georgia, Athens
Willie M. May, Wendell Phillips High School, Chicago, Illinois
Philip L. Miller, Carnegie-Mellon University, Pittsburgh, Pennsylvania
Judith T. Sutcliffe, St. Mark's School of Texas, Dallas

Acknowledgments

The College Board wishes to thank all the individuals and organizations that contributed to *Academic Preparation in Mathematics*. In addition to those who served on the Mathematical Sciences Advisory Committee and the Council on Academic Affairs, explicit acknowledgment should be accorded to Yola Coffeen, Robert Orrill, Mary Carroll Scott, and Carol J. Meyer. Without the leadership of Adrienne Y. Bailey, vice president for Academic Affairs at the College Board, this book would not have assumed its present form. Although none of these people is individually responsible for the contents of the book, the Educational EQuality Project owes much to their efforts.

James Herbert, General Editor

The College Board is a nonprofit membership organization that provides tests and other educational services for students, schools, and colleges. The membership is composed of more than 2,500 colleges, schools, school systems, and education associations. Representatives of the members serve on the Board of Trustees and advisory councils and committees that consider the College Board's programs and participate in the determination of its policies and activities.

The Educational EQuality Project is a 10-year effort of the College Board to strengthen the academic quality of secondary education and to ensure equality of opportunity for postsecondary education for all students. Begun in 1980, the project is under the direction of the Board's Office of Academic Affairs.

For more information about the Educational EQuality Project and inquiries about this report, write to the Office of Academic Affairs, The College Board, 45 Columbus Avenue, New York, New York 10023-6917.

To Our Fellow Teachers of Mathematics

Those of us who have made a commitment to teaching mathematics know that mathematics is a living, growing subject. We also know that teaching mathematics in such a way as to reflect the vitality of the subject itself imposes a tremendous responsibility, which sometimes we wish we did not have to face. The gap between what our students know and what they need to know to use mathematics intelligently in later life can be discouraging when we look at it too closely.

What keeps us going is the recognition that, for all the frustrations of teaching mathematics today, the subject we teach still has the power to ignite young minds. What teacher does not feel a thrill when a student suddenly sees the sense behind what was, until then, a mechanical process, or when a group of students discover that the special cases they have been looking at fall into a beautiful pattern that can be generalized?

Many of our students—perhaps most of them—will enter college. We try to offer them the preparation in mathematics that they will need to succeed not only in higher education but also in their later endeavors. Our task is complicated by the fact that mathematics as a discipline is changing rapidly. Furthermore, many fields of study are demanding more and better preparation in mathematics for those who enter them. We cannot predict with much assurance which of our students will need advanced preparation in mathematics and which will not. We cannot even predict with certainty which of our students will pursue higher education.

The only solution appears to be to offer all our students who show any potential for college the kind of broad preparation in mathematics that will equip them to use mathematics productively in daily life and that will enable them to learn later the specific skills they might need in their careers. As we encounter students who intend to pursue additional study in mathematics, we can provide programs that both accelerate and enrich their preparation for college. But we should always keep an eye on those students who are shutting mathematics out of their life and avoiding the challenge it presents. When we hold their interest and give them a chance to succeed in learning some aspect of mathematics, we are opening a door for them that can lead to opportunities they might never have had.

There are experts on every corner today who are quick to tell anyone who will listen how teachers are failing. Those of us who have continued on the job know how hard it has become in the last few years—and how few good words are spoken in support of teachers. The older teachers among us have heard most of these complaints before, but the tone seems to have become more caustic lately.

All of us know that teaching is a difficult job, but most of us also know that we are *not* failing at it. As a matter of fact, we are succeeding for the most part. We also know that, with more support and assistance, we could be doing better.

The purpose of this book is to provide some assistance to teachers who would like to help students enter college with the kind of preparation in mathematics they need to be successful. In this book, we offer some ideas for our fellow mathematics teachers to consider. We have tried to specify the outcomes that should result from the study of mathematics and have offered suggestions for achieving these outcomes. We view our ideas as the beginning of a dialogue that will help renew the teaching of mathematics in schools. We hope that teachers will be encouraged to collaborate with their colleagues in attempting to put these suggestions into practice. We have tried to lay out alternatives, but we have not given a detailed blueprint for change. No one can tell in advance whether or not a proposal about teaching will work in a particular classroom with a particular teacher. Someone else's experience may be helpful, but there is no recourse for teachers who want change except to try ideas themselves.

The last big wave of curriculum reform in mathematics, in the 1950s and 1960s, came down from above. Teachers were not truly involved in the process of change; by and large, they were expected to do what someone else had decided. That kind of change does not always get implemented in the classroom and, when it does, it is short-lived.

If the outcomes of college preparation in mathematics are to change so that more students have more opportunity to study more mathematics more extensively, that change must begin with the mathematics teacher—with you. The curriculum, as it ultimately reaches the student, is fashioned by the teacher—no matter who writes the textbook or constructs the standardized test. The mathematics teacher is not, and should not be, expected to develop a complete curriculum single-handedly. But he or she does play the central role in transforming ideas about what should be taught into ideas that are taught. We invite you to join us in the enterprise of reshaping college preparatory mathematics in your school.

We hope this book will be useful in discussions that contribute to that enterprise. In Chapter 1 we sketch the relation of this book to the College Board's Educational EQuality Project and to the project's earlier publication *Academic Preparation for College: What Students Need to Know and Be Able to Do*. In Chapter 2 we provide greater detail concerning the learning outcomes needed by all college entrants. Our suggestions concerning curriculum for achieving those outcomes are outlined in Chapter 3, and in Chapter 4 we present some vignettes that might suggest ways of making curriculum effective in classroom instruction. Chapter 5 relates teaching and learning in mathematics to other academic competencies such as reading, writing, and reasoning. In Chapter 6 we suggest some areas where we believe further discussion and exploration are needed.

Further discussion is the heart of our message in this book. By working together teachers of mathematics can achieve significant improvement in how our subject is taught and learned. So often mathematics has been seen as arbitrary and irrelevant. We want it to be meaningful, memorable, and applicable. So often mathematics has been a barrier to college entrance. We want it to become a bridge.

Mathematical Sciences Advisory Committee

I. Beyond the Green Book

Identifying the academic preparation needed for college is a first step toward providing that preparation for all students who might aspire to higher education. But the real work of actually achieving these learning outcomes lies ahead.[1]

This book is a sequel to *Academic Preparation for College: What Students Need to Know and Be Able to Do*, which was published in 1983 by the College Board's Educational EQuality Project. Now widely known as the Green Book, *Academic Preparation for College* outlined the knowledge and skills students need in order to have a fair chance at succeeding in college. It summarized the combined judgments of hundreds of educators in every part of the country. The Green Book sketched learning outcomes that could serve as goals for high school curricula in six Basic Academic Subjects: English, the arts, mathematics, science, social studies, and foreign languages. It also identified six Basic Academic Competencies on which depend, and which are further developed by, work in these subjects. Those competencies are reading, writing, speaking and listening, mathematics, reasoning, and studying. The Green Book also called attention to additional competencies in using computers and observing, whose value to the college entrant increasingly is being appreciated.

With this book we take a step beyond *Academic Preparation for College*. The Green Book simply outlined desired results of high school education—the learning all students need to be adequately prepared for college. It contained no specific suggestions about how to achieve those results. Those of us working with the Educational EQuality Project strongly believed—and still believe—that ulti-

---

1. The College Board, *Academic Preparation for College: What Students Need to Know and Be Able to Do* (New York: The College Board, 1983), p. 31.

mately curriculum and instruction are matters of local expertise and responsibility. Building consensus on goals, while leaving flexible the means to achieve them, makes the most of educators' ability to respond appropriately and creatively to conditions in their own schools. Nevertheless, teachers and administrators, particularly those closely associated with the EQuality project, often have asked how the outcomes sketched in the Green Book might be translated into actual curricula and instructional practices—how they can get on with the "real work" of education. These requests in part seek suggestions about how the Green Book goals might be achieved; perhaps to an even greater extent they express a desire to get a fuller picture of those very briefly stated goals. Educators prefer to think realistically, in terms of courses and lessons. Discussion of proposals such as those in the Green Book proceeds more readily when goals are filled out and cast into the practical language of possible courses of action.

To respond to these requests for greater detail, and to encourage further nationwide discussion about what should be happening in our high school classrooms, teachers working with the Educational EQuality Project have prepared this book and five like it, one in each of the Basic Academic Subjects. By providing suggestions about how the outcomes described in *Academic Preparation for College* might be achieved, we hope to add more color and texture to the sketches in that earlier publication. We do not mean these suggestions to be prescriptive or definitive, but to spark more detailed discussion and ongoing dialogue among our fellow teachers who have the front-line responsibility for ensuring that all students are prepared adequately for college. We also intend this book and its companions for guidance counselors, principals, superintendents, and other officials who must understand the work of high school teachers if they are better to support and cooperate with them.

Students at Risk, Nation at Risk

Academic Preparation for College was the result of an extensive grassroots effort involving hundreds of educators in every part of the country. However, it was not published in a vacuum. Since the beginning of this decade, many blue-ribbon commissions and studies also have focused attention on secondary education. The concerns of these reports have been twofold. One, the reports note a perceptible decline in the academic attainments of students who graduate from high school, as indicated by such means as standardized test scores and comments from employers; two, the reports reflect a widespread worry that, unless students are better educated, our national welfare will be in jeopardy. *A Nation at Risk* made this point quite bluntly:

> Our Nation is at risk. Our once unchallenged preeminence in commerce, industry, science, and technological innovation is being overtaken by competitors throughout the world. . . . The educational foundations of our society are presently being eroded by a rising tide of mediocrity that threatens our very future as a Nation and a people.[2]

The Educational EQuality Project, an effort of the College Board throughout the decade of the 1980s to improve both the quality of preparation for college and the equality of access to it, sees another aspect of risk: if our nation is at risk because of the level of students' educational attainment, then we must be more concerned with those students who have been most at risk.

Overall, the predominance of the young in our society is ending. In 1981, as the EQuality project was getting under way, about 41 percent of our country's population was under 25 years old and 26 percent was 50 years old or older. By the year 2000, however, the balance will have shifted to 34 percent and 28 percent, respectively. But these figures do not tell the whole story, especially for those of us working in the schools. Among certain groups, youth is a growing segment of the population. For example, in 1981, 71 percent of black and 75 percent of Hispanic households had children 18 years old or younger. In comparison, only 52 percent of all white households had children in that age category. At the beginning of the 1980s, children from minority groups already made up more than

---

2. National Commission on Excellence in Education, *A Nation at Risk* (Washington, D.C.: U.S. Government Printing Office, 1983), p. 5.

25 percent of all public school students.[3] Clearly, concern for improving the educational attainments of all students increasingly must involve concern for students from such groups of historically disadvantaged Americans.

How well will such young people be educated? In a careful and thoughtful study of schools, John Goodlad found that "consistent with the findings of virtually every study that has considered the distribution of poor and minority students . . . minority students were found in disproportionately large percentages in the low track classes of the multiracial samples [of the schools studied]."[4] The teaching and learning that occur in many such courses can be disappointing in comparison to that occurring in other courses. Goodlad reported that in many such courses very little is expected, and very little is attempted.[5]

When such students are at risk, the nation itself is at risk, not only economically but morally. That is why this book and its five companions offer suggestions that will be useful in achieving academic excellence for *all* students. We have attempted to take into account that the resources of some high schools may be limited and that some beginning high school students may not be well prepared. We have tried to identify ways to keep open the option of preparing adequately for college as late as possible in the high school years. These books are intended for work with the broad spectrum of high school students—not just a few students and not only those currently in the "academic track." We are convinced that many more students can—and, in justice, should—profit from higher education and therefore from adequate academic preparation.

Moreover, many more students actually enroll in postsecondary education than currently follow the "academic track" in high

3. Ernest L. Boyer, *High School* (New York: Harper & Row, 1983), pp. 4-5. U.S. Department of Education, National Center for Education Statistics, *Digest of Education Statistics: 1982* (Washington, D.C.: U.S. Government Printing Office, 1982), p. 43.

4. John Goodlad, *A Place Called School* (New York: McGraw-Hill, 1984), p. 156.
5. Ibid., p. 159.

school. Further, discussions with employers have emphasized that many of the same academic competencies needed by college-bound students also are needed by high school students going directly into the world of work. Consequently, the Educational EQuality Project, as its name indicates, hopes to contribute to achieving a democratic excellence in our high schools.

The Classroom: At the Beginning as Well as the End of Improvement

A small book such as this one, intended only to stimulate dialogue about what happens in the classroom, cannot address all the problems of secondary education. On the other hand, we believe that teachers and the actual work of education—that is to say, curriculum and instruction—should be a more prominent part of the nationwide discussion about improving secondary education.

A 1984 report by the Education Commission of the States found that 44 states either had raised high school graduation requirements or had such changes pending. Twenty-seven states had enacted new policies dealing with instructional time, such as new extracurricular policies and reduced class sizes.[6] This activity reflects the momentum for and concern about reform that has been generated recently. It demonstrates a widespread recognition that critiques of education without concrete proposals for change will not serve the cause of improvement. But what will such changes actually mean in the classroom? New course requirements do not necessarily deal with the academic quality of the courses used to fulfill those requirements. Certain other kinds of requirements can force instruction to focus on the rote acquisition of information to the exclusion of fuller intellectual development. Manifestly, juggling of requirements and courses without attention to what needs to occur between teachers and students inside the classroom will not automatically

---

6. *Action in the States: Progress toward Education Renewal*, A Report by the Task Force on Education for Economic Growth (Denver, Colorado: Education Commission of the States, 1984), p. 27.

produce better prepared students. One proponent of reform, Ernest Boyer, has noted that there is danger in the prevalence of "quick-fix" responses to the call for improvement. "The depth of discussion about the curriculum . . . has not led to a serious and creative look at the nature of the curriculum. . . . states [have not asked] what we ought to be teaching."[7]

Such questioning and discussion is overdue. Clearly, many improvements in secondary education require action outside the classroom and the school. Equally clearly, even this action should be geared to a richer, more developed understanding of what is needed in the classroom. By publishing these books we hope to add balance to the national debate about improving high school education. Our point is not only that it is what happens between teachers and students in the classroom that makes the difference. Our point is also that what teachers and students do in classrooms must be thoughtfully considered before many kinds of changes, even exterior changes, are made in the name of educational improvement.

From Deficit to Development

What we can do in the classroom is limited, of course, by other factors. Students must be there to benefit from what happens in class. Teachers know firsthand that far too many young people of high school age are no longer even enrolled. Nationally, the drop-out rate in 1980 among the high school population aged 14 to 34 was 13 percent. It was higher among low-income and minority students. Nearly 1 out of 10 high schools had a drop-out rate of over 20 percent.[8]

Even when students stay in high school, we know that they do not always have access to the academic preparation they need.

---

7. In Thomas Toch, "For School Reform's Top Salesmen, It's Been Some Year," *Education Week*, June 6, 1984, p. 33.

8. National Center for Education Statistics, *Digest of Education Statistics: 1982*, p. 68. Donald A. Rock, et al., "Factors Associated with Test Score Decline: Briefing Paper" (Princeton, New Jersey: Educational Testing Service, December 1984), p. 4.

Many do not take enough of the right kinds of courses. In 1980, in almost half of all high schools, a majority of the students in each of those schools was enrolled in the "general" curriculum. Nationwide, only 38 percent of high school seniors had been in an academic program; another 36 percent had been in a general program; and 24 percent had followed a vocational/technical track. Only 39 percent of these seniors had enrolled for three or more years in history or social studies; only 33 percent had taken three or more years of mathematics; barely 22 percent had taken three or more years of science; and less than 8 percent of these students had studied Spanish, French, or German for three or more years.[9]

Better than anyone else, teachers know that, even when students are in high school and are enrolled in the needed academic courses, they must attend class regularly. Yet some school systems report daily absence rates as high as 20 percent. When 1 out of 5 students enrolled in a course is not actually there, it is difficult even to begin carrying out a sustained, coherent program of academic preparation.

As teachers we know that such problems cannot be solved solely by our efforts in the classroom. In a world of disrupted family and community structures; economic hardship; and rising teenage pregnancy, alcoholism, and suicide, it would be foolish to believe that attention to curriculum and instruction can remedy all the problems leading to students' leaving high school, taking the wrong courses, and missing classes. Nonetheless, what happens in the high school classroom—once students are there—is important in preparing students not only for further education but for life.

Moreover, as teachers, we also hope that what happens in the classroom at least can help students stick with their academic work. Students may be increasingly receptive to this view. In 1980 more than 70 percent of high school seniors wanted greater academic emphasis in their schools; this was true of students in all curricula. Mortimer Adler may have described a great opportunity:

> There is little joy in most of the learning they [students] are now compelled to do. Too much of it is make-believe, in which neither

---

9. National Center for Education Statistics, *Digest of Education Statistics: 1982*, p. 70.

teacher nor pupil can take a lively interest. Without some joy in learning—a joy that arises from hard work well done and from the participation of one's mind in a common task—basic schooling cannot initiate the young into the life of learning, let alone give them the skill and the incentive to engage in it further.[10]

Genuine academic work can contribute to student motivation and persistence. Goodlad's study argues strongly that the widespread focus on the rote mechanics of a subject is a surefire way to distance students from it or to ensure that they do not comprehend all that they are capable of understanding. Students need to be encouraged to become inquiring, involved learners. It is worth trying to find more and better ways to engage them actively in the learning process, to build on their strengths and enthusiasms. Consequently, the approaches suggested in these books try to shift attention from chronicling what students do not know toward developing the full intellectual attainments of which they are capable and which they will need in college.

Dimensions for a Continuing Dialogue

This book and its five companions were prepared during 1984 and 1985 under the aegis of the College Board's Academic Advisory Committees. Although each committee focused on the particular issues facing its subject, the committees had common purposes and common approaches. Those purposes and approaches may help give shape to the discussion that this book and its companions hope to stimulate.

Each committee sought the assistance of distinguished writers and consultants. The committees considered suggestions made in the dialogues that preceded and contributed to *Academic Preparation for College* and called on guest colleagues for further suggestions and insights. Each committee tried to take account of the best available thinking and research but did not merely pass along the results of research or experience. Each deliberated about those

---

10. Mortimer J. Adler, *The Paideia Proposal: An Educational Manifesto* (New York: Macmillan Publishing Company, 1982), p. 32.

findings and then tried to suggest approaches that had actually worked to achieve learning outcomes described in *Academic Preparation for College*. The suggestions in these books are based to a great extent on actual, successful high school programs.

These books focus not only on achieving the outcomes for a particular subject described in the Green Book but also on how study of that subject can develop the Basic Academic Competencies. The learning special to each subject has a central role to play in preparing students for successful work in college. That role ought not to be neglected in a rush to equip students with more general skills. It is learning in a subject that can engage a student's interest, activity, and commitment. Students do, after all, read about *something*, write about *something*, reason about *something*. We thought it important to suggest that developing the Basic Academic Competencies does not replace, but can result from, mastering the unique knowledge and skills of each Basic Academic Subject. Students, particularly hungry and undernourished ones, should not be asked to master the use of the fork, knife, and spoon without being served an appetizing, full, and nourishing meal.

In preparing the book for each subject, we also tried to keep in mind the connections among the Basic Academic Subjects. For example, the teaching of English and the other languages should build on students' natural linguistic appetite and development— and this lesson may apply to the teaching of other subjects as well. The teaching of history with emphasis on the full range of human experience, as approached through both social and global history, bears on the issue of broadening the "canon" of respected works in literature and the arts. The teaching of social studies, like the teaching of science, involves mathematics not only as a tool but as a mode of thought. There is much more to make explicit and to explore in such connections among the Basic Academic Subjects. Teachers may teach in separate departments, but students' thought is probably not divided in the same way.

Finally, the suggestions advanced here generally identify alternate ways of working toward the same outcomes. We wanted very much to avoid any hint that there is one and only one way to achieve the outcomes described in *Academic Preparation for College*. There are many good ways of achieving the desired results, each one good in its own way and in particular circumstances. By describing alternate approaches, we hope to encourage readers of this book to analyze and recombine alternatives and to create the most appropriate and effective approaches, given their own particular situations.

We think that this book and its five companion volumes can be useful to many people. Individual teachers may discover suggestions that will spur their own thought about what might be done in the classroom; small groups of teachers may find the book useful in reconsidering the mathematics program in their high school. It also may provide a takeoff point for in-service sessions. Teachers in several subjects might use it and its companions to explore concerns, such as the Basic Academic Competencies, that range across the high school curriculum. Principals may find these volumes useful in refreshing the knowledge and understanding on which their own instructional leadership is based.

We also hope that these books will prove useful to committees of teachers and officials in local school districts and at the state level who are examining the high school curriculum established in their jurisdictions. Public officials whose decisions directly or indirectly affect the conditions under which teaching and learning occur may find in the books an instructive glimpse of the kinds of things that should be made possible in the classroom.

Colleges and universities may find in all six books occasion to consider not only how they are preparing future teachers, but also whether their own curricula will be suited to students receiving the kinds of preparation these books suggest. But our greatest hope is that this book and its companions will be used as reference points for dialogues between high school and college teachers. It was from such dialogues that *Academic Preparation for College* emerged. We believe that further discussions of this sort can provide a wellspring of insight and energy to move beyond the Green Book toward actually providing the knowledge and skills all students need to be successful in college.

We understand the limitations of the suggestions presented here. Concerning what happens in the classroom, many teachers, researchers, and professional associations can speak with far greater depth and detail than is possible in the pages that follow. This book aspires only to get that conversation going, particularly across the boundaries that usually divide those concerned about education, and especially as it concerns the students who often are least well served. Curriculum, teaching, and learning are far too central to be omitted from the discussion about improving education.

II. Preparation and Outcomes

In this chapter, we consider the preparation in mathematics that students need for college. We begin with an overview of the general skills—the Basic Academic Competencies—that students need for effective work in mathematics and that are developed by its study. Then we present the specific skills and knowledge that should be outcomes of the study of mathematics in high school. Both these sections elaborate the outcomes sketched in *Academic Preparation for College*. The chapter ends with a discussion of what students need to know and be able to do in mathematics on entry to high school.

Developing the Basic Academic Competencies

The Basic Academic Competencies are reading, writing, speaking and listening, mathematics, reasoning, and studying. The Green Book suggests that this list soon may come to include observing and computer competency. Mathematics is unique in that it is both a Basic Academic Competency and a Basic Academic Subject. This dual role reflects the dual position that mathematics historically has occupied in the school curriculum. It is a "tool" subject that provides students with a family of specialized languages and associated skills for dealing with quantitative problems from any field of endeavor, and it is simultaneously one of the liberal arts whose mastery has marked an educated person since the time of Plato. Students need to be exposed to both faces of mathematics. They need to see that mathematics as an academic subject both depends on and strengthens mathematics as an academic competency; the content of the two aspects of mathematics should be in harmony.

Moreover, the study of mathematics both depends on and strengthens each of the other Basic Academic Competencies. This section describes each competency and, to promote understanding, frequently includes illustrative suggestions of how the competency can be developed through the study of mathematics in high school. A more nearly complete discussion of the interrelation between mathematics and other basic competencies will be provided in Chapter 5.

Reading

The study of mathematics can contribute to a student's general competence in reading. For example, the skills of deductive inference developed through the study of logic and axiomatic systems can enhance students' ability to read analytically. In the mathematics class, a student can learn the skills of close, critical reading that are vital in such fields as law and medicine. To read and comprehend a symbolic mathematical expression requires that each symbol be read and given an unambiguous interpretation, that key symbols be given special attention, that the relations between symbols be understood, and that relevant rules for operating with the symbols be recalled. The reader must be able to examine such an expression bit by bit, if necessary, to be sure that one small symbol—an exponent, for example—is not misread. At the same time, the reader should be able to "take in the expression at a glance," so to speak, to decide whether or not it is of a familiar type to which standard procedures apply. Depending on how it is taught, mathematics can offer students many opportunities to develop their reading skills.

Writing

The study of mathematics can help develop a student's competence in writing. For example, when Jean writes up her solution to a problem so that Susan, who has not been able to solve it, can follow Jean's solution without having to ask further questions, Jean learns something about anticipating the needs of one's readers. When students document a computer program they have written, they must avoid omitting key steps on the one hand and getting bogged down in minor points on the other. Students who are proficient in describing the algorithms they develop have acquired important skills of selecting and restructuring their ideas to communicate a coherent message.

Teachers who have students keep journals to record class discussions, problem solutions, and their own observations about a mathematical topic find that the discipline of maintaining a written record helps students remember key ideas and see relationships among the various topics in the course. When asked to write an explanation of a topic, students who have kept a journal write more thoughtful papers because they have learned the advantage of writing one's preliminary thoughts before attempting to write a finished essay.

Speaking and Listening

Students who are given the opportunity to speak to a small group of their classmates or to the entire class on their investigation of a mathematical problem not only enhance their understanding of the mathematics involved but also develop skill in organizing and presenting ideas orally. Students can develop skills in critical listening if they have opportunities to evaluate the coherence and completeness of a spoken mathematical argument presented by their classmates or by the teacher. Teachers who have abandoned the role of "supreme answer source" find that students listen more carefully to each other's questions and provide clearer, more incisive answers.

Reasoning

The study of high school mathematics requires that students understand something of both inductive and deductive reasoning. Skills of deductive reasoning are both required in and enhanced by the process of proving theorems, but formal proofs are by no means necessary for the student to begin to appreciate the power of deductive reasoning in mathematics. Even in introductory courses, students can be encouraged to state reasons for the key steps they are taking in solving a mathematical problem, to suggest why they think a particular rule works for any number, and to support their arguments by appealing to results they have established previously.

Students also need to understand how inductive reasoning works in mathematics through assignments in which they investigate a pattern—such as the sequence of sums $1, 3 + 5, 7 + 9 + 11, 13 + 15 + 17 + 19$, and so on; or the number of regions into which $1, 2, 3, 4$, or more chords divide the interior of a circle—and then conjecture the general law suggested by these cases. Students who are given opportunities to test their conjectures learn that some conjectures may be wrong and that ultimately all valid conjectures must be proved if they are to be used in further deductive arguments.

Studying

Although the study of mathematics often is thought of as a solitary pursuit, students engaging in collaborative study experience mathematics as the cooperative activity it so often is in business and industry. When students work in pairs on an assignment, they teach and learn from each other. Examination questions can be assigned to students in groups; they gain valuable experience in using reading, writing, listening, and speaking skills as they work together to prepare a collective response. Tasks that require students to locate information on mathematics in periodicals and books other than textbooks, or in computerized data bases, can help students develop their research skills as well as learn how mathematics is used outside school.

Mathematics

The study of mathematics in high school builds upon and develops the broad intellectual skills in mathematics that students need in all fields of college study. So that the reader can compare the specific outcomes of the study of high school mathematics that are discussed in the next section with the general competencies in mathematics that should be developed and used in all high school courses, we repeat the list of Basic Academic Competencies in mathematics given in *Academic Preparation for College*.

Library - St. Joseph's College
222 Clinton Avenue
Brooklyn, N.Y. 11205

- The ability to perform, with reasonable accuracy, the computations of addition, subtraction, multiplication, and division using natural numbers, fractions, decimals, and integers.
- The ability to make and use measurements in both traditional and metric units.
- The ability to use effectively the mathematics of:
  - integers, fractions, and decimals
  - ratios, proportions, and percentages
  - roots and powers
  - algebra
  - geometry
- The ability to make estimates and approximations, and to judge the reasonableness of a result.
- The ability to formulate and solve a problem in mathematical terms.
- The ability to select and use appropriate approaches and tools in solving problems (mental computation, trial and error, paper-and-pencil techniques, calculator, and computer).
- The ability to use elementary concepts of probability and statistics.

Outcomes of the Study of Mathematics in High School

The central role of the study of mathematics in preparing students for college has long been acknowledged by society, but the need for high levels of preparation has not always been appreciated. Neither individual students nor society itself should accept a level of mathematical literacy that restricts opportunities for career choices and personal advancement. In the 1980s and 1990s students preparing for college will need more and better preparation in mathematics than has been required of them in the past. In the coming decade, all college entrants will need to achieve proficiency in the following basic mathematical skills:

- The ability to apply mathematical techniques in the solution of real-life problems and to recognize when to apply those techniques.

- Familiarity with the language, notation, and deductive nature of mathematics and the ability to express quantitative ideas with precision.
- The ability to use computers and calculators.
- Familiarity with the basic concepts of statistics and statistical reasoning.
- Knowledge in considerable depth and detail of algebra, geometry, and functions.

These outcomes are examined below in greater detail. The first two are outcomes that should result from all mathematics instruction and are discussed under the heading "Applications and Language." They have been placed first not only because of their central importance in the school mathematics curriculum but also because they build directly upon mathematics as a Basic Academic Competency, particularly with respect to the use of mathematics in solving problems.

The other outcomes are discussed under the headings of specific mathematical topics. For each topic, we offer some comments on its importance and how it contributes to the student's education. We list specific proficiencies that every college entrant will need. Then we discuss the additional preparation in that topic that will be required by those students who expect to major in such heavily mathematical fields as science or engineering, or by those who expect to take advanced courses in mathematics or computer science.

Applications and Language

High school students need to see mathematics applied to problems outside the classroom and how people in various careers use mathematics. No one can anticipate the particular applications that a student will need as an adult, but every student can be given practice in applying a variety of mathematical ideas and methods. For example, the design of a package for a commercial product would illustrate the application of ideas from algebra, geometry, and statistics. An examination of the process of radioactive decay would illustrate how an exponential function is used; students can graph the function and examine its properties. A wide variety of such applications is now available in resource material designed for use by high school mathematics teachers.[11,12]

The effective application of mathematics requires skill in formulating and solving problems. That skill, in turn, requires a mastery of basic mathematical knowledge and processes, explicit instruction in various techniques, and extensive practice. Instruction in problem solving must include attention to specific techniques that problem solvers use, such as drawing a figure, listing and tabulating key facts, and examining special cases. Students need to learn not only how to use such techniques but also how to recognize when they might be helpful.

All high school students preparing for college should learn to use the language and notation of mathematics effectively. For example, they should be able to distinguish between a bisector of a line segment and the midpoint of that segment, and they should be able

to interpret $\sum_{n=1}^{3} 3^{n+1}$ as equivalent to $3^2 + 3^3 + 3^4$. They should be

able to use appropriately such terms as *polynomial, angle,* and *function* when they write or talk about mathematics.

Students should have frequent opportunities to see mathematical rules develop in a logical sequence. Although Euclidean geometry can provide a fertile ground for gaining an appreciation of the axiomatic aspect of mathematics, it should not be the exclusive source of experiences with deductive reasoning. Experiences in pattern recognition and inductive reasoning should be organized so

---

11. Joint Committee of the Mathematical Association of America and the National Council of Teachers of Mathematics, *A Sourcebook of Applications of School Mathematics* (Reston, Virginia: National Council of Teachers of Mathematics, 1980).

12. Sidney Sharron and Robert E. Reys (eds.), *Applications in School Mathematics* (Reston, Virginia: National Council of Teachers of Mathematics, 1979).

as to lead to the development of mathematical laws and concepts.

Beyond these general outcomes of applications and language, college entrants will need the preparation in mathematics outlined in the following sections.

Computing

Computers are affecting how mathematics is taught in much the same way as they are affecting the teaching of other subjects. As computers become increasingly common in schools, computer competency and computer programming ability should become natural outcomes of, rather than adjuncts to, mathematics programs. Computers and hand-held calculators should be used in mathematics classes to facilitate computations, and some instruction in their intelligent use is necessary.

Computers also are affecting the subject of mathematics in a unique way. Few people realize that computer science is changing the shape of mathematics as a discipline and consequently will affect the mathematics taught in schools and colleges. Computers are making mathematics into a subject that requires much less attention to the mastery of routine procedures and much more to how computer power can be used to explore mathematical ideas.

Consequently, every student entering college in the 1990s will need to be able to use computer programs to perform arithmetic and algebraic calculations, produce graphs and figures, and interact with data bases and networks.[13]

The availability of the hand calculator for performing arithmetic operations has made obsolete the necessity for students to learn elaborate paper-and-pencil algorithms for calculation. Learning the long-division algorithm and the "mark-off-two-places" algorithm for taking square roots is a waste of time in an age when anyone required to perform such computations outside of school has easy access to a calculator. Instead, more than ever before, students

---

13. James T. Fey (ed.), *Computing and Mathematics: The Impact on Secondary School Curricula* (Reston, Virginia: National Council of Teachers of Mathematics, 1984), pp. 5-6.

must be able to estimate the magnitude of a result so that they can be sensitive to possible errors in a machine calculation, and they need skills in mental arithmetic so that they can check a calculation to determine where an error might have occurred.

Using calculators and computers in doing mathematics can help students organize their thinking as they make explicit the organization and procedures needed to solve a particular problem. The ideas of breaking a problem into parts, formulating procedures for each part, and debugging faulty procedures should be familiar to students as techniques that also can be applied to problems outside mathematics. All students should have some experience in devising algorithms to solve mathematical problems and should be introduced to the use of prepared computer programs for the implementation of such algorithms. Students need practice in using computers to simulate the results of games and experiments so that they can learn to apply mathematical models to realistic sets of data.

In sum, college entrants will need:

- Familiarity with computer programming and the use of prepared computer programs in mathematics.
- The ability to use mental computation and estimation to evaluate calculator and computer results.
- Familiarity with the methods used to solve mathematical problems when calculators or computers are the tools.

Those students who are expecting to major in science, engineering, or business in college, or who expect to take advanced courses in mathematics or computer science, will need additional preparation in computing. They should be able to do more than simply use programs and should have a general familiarity with how computer methods operate. They will need the following more extensive mathematical proficiency.

- The ability to write computer programs to solve a variety of mathematical problems.
- Familiarity with the methodology of developing computer programs and with the considerations of design, structure, and style that are an important part of this methodology.

Statistics

Students need to be given a foundation in the mathematical concepts that underlie applications of statistical reasoning and the theory of probability. Perhaps no other topics in mathematics surround and affect our lives as much and yet are so poorly understood as statistics and probability. We encounter them in daily life when we hear about the reported chances of rain, the test marketing of a new product, or the meaning of an economic indicator. Students who enter college in the 1990s will find that statistics are being employed to solve important problems in most of the subject fields they study. Increasingly, college students majoring in such diverse and seemingly nonmathematical fields as forestry, criminal justice, and food science are finding that a background in statistics and probability is a necessity. Yet in most high schools the mathematics curriculum gives only minimal attention to statistics.

Students entering college should have had experience in gathering and interpreting statistical data and should be sensitive to various misuses of statistics. They should know how to take data and represent them in various graphical forms, depending on the nature of the data and the purpose of the representation. They should be able to summarize data in various ways, describing the shape of a distribution and noting the range and central tendency. They should be able to reason that if, for example, 34 out of a sample of 103 students in the school have their own cars, perhaps about one third of all the students do. Further, they should understand that how the sample was drawn will affect the confidence with which one can generalize about the whole school.

The treatment of statistics in high school should be based on problems drawn from daily life. The availability of computers makes feasible the use of realistic data and allows students to see at a glance how different mathematical treatments of the data can lead to different results. Students do not need to spend time in calculating complicated statistics or in deriving theorems from axioms of probability. Instead, they can gain an intuitive understanding of how statistical methods work by applying them to realistic problems and observing how changes in a problem affect the statistics. Although the inclusion of statistics in the secondary mathematics curriculum has been advocated for decades, the availability of calculators, computers, and useful teaching materials at last has made it possible for meaningful activities involving applied statistics to be conducted by the teacher in any high school mathematics classroom.[14,15]

College entrants will need:

- The ability to gather and interpret data and to represent them graphically.
- The ability to apply techniques for summarizing data, using such statistical concepts as average, median, and mode.
- Familiarity with techniques of statistical reasoning and common misuses of statistics.

If students intend to pursue majors or careers requiring advanced mathematics courses in college, they will need more preparation in statistics. They should have opportunities in high school, if not before, to use the computer to model experimental situations, analyzing the mathematics needed in the model. They should understand and be able to use such techniques as random number generators to study the simulated behavior of individual organisms or social systems. They should be able to apply counting techniques and concepts relating to permutations and combinations to solve realistic problems. They also will need more extensive grounding in probability theory, although a rigorously deductive approach should be avoided. The additional preparation they need includes the following:

- Understanding of simulation techniques used to model experimental situations.
- Knowledge of elementary concepts of probability needed in the study and understanding of statistics.

Algebra

For many years, algebra has been the field of mathematics most strongly associated with college entrance. First-year algebra begins

14. Albert P. Shulte and James R. Smart (eds.), *Teaching Statistics and Probability* (Reston, Virginia: National Council of Teachers of Mathematics, 1981 Yearbook).

15. Judith M. Tanur, Frederick Mosteller, William H. Kruskal, Richard F. Link, Richard S. Pieters, and Gerald R. Rising (eds.), *Statistics: A Guide to the Unknown* (San Francisco: Holden-Day, 1972).

the college-entrance sequence of mathematical studies in many high schools. Algebra is often the first (and perhaps the only) required course in mathematics for students entering various nontechnical fields. Professors of calculus, physics, chemistry, and biology constantly complain about ineptitude in algebra when they view their students' preparation in mathematics. In recent years, many colleges and universities have instituted tests of skill in algebraic manipulation as a device for screening students before placing them in appropriate mathematics courses. The student who enters college with rusty or nonexistent algebraic skills is likely to be headed for a remedial program.

But just as the hand calculator has made obsolete various requirements for computational skill in arithmetic, so the impending appearance of calculators that perform symbolic manipulations, such as solving quadratic equations and adding rational expressions, promises to reshape the requirements for computational skill in algebra. Computer programs are already on the market that can perform all of the symbolic manipulations required for high school algebra.

As machines that can do symbolic manipulations become available in classrooms, teachers will be able to direct more of their students' attention to the meaning of the various operations of algebra. More attention can be given to how algebraic relations can be used to model real situations and how such relations are represented symbolically (and graphically). Students still will need to know how to solve equations and how to transform algebraic expressions in various ways, but they will not need the level of manipulative skill once expected of them. For example, instead of spending weeks learning how to solve quadratic equations by laborious paper-and-pencil approaches involving factoring or completing the square, students can use the quadratic formula to explore various important applications such as the motion of projectiles or nonlinear cost-and-demand functions. They also can examine how the behavior of the roots depends on the coefficients and can use various techniques to solve polynomial equations by approximation. In this way, they can develop a fuller understanding of what it means to solve an equation.

The new technology can help teachers alter the usual approach to instruction in algebra that places heavy emphasis on the mastery of technique and seldom addresses applications. As Fey and his colleagues have observed, "With the aid of computers that perform algorithmic manipulations, the conventional order and priority of topics could be inverted. The algebra experience could begin with, and focus on, the processes of expressing and interpreting quantitative relations in symbolic (and graphic) form, with far less attention paid to the techniques of symbol manipulation."[16]

College entrants will need:

- Skill in solving equations and inequalities.
- Skill in operations with real numbers.
- Skill in simplifying algebraic expressions, including simple rational and radical expressions.
- Familiarity with permutations, combinations, simple counting problems, and the binomial theorem.

Those students who will take advanced mathematics in college will need additional preparation in high school, including:

- Skill in solving trigonometric, exponential, and logarithmic equations.
- Skill in operations with complex numbers.
- Familiarity with arithmetic and geometric series and with proofs by mathematical induction.
- Familiarity with simple matrix operations and their relation to systems of linear equations.

Geometry

Although classical geometry plays a limited role in contemporary mathematics, the study of geometry is of fundamental importance in developing students' intuitions about the mathematics that they will study later in a more abstract form. Students need to develop an understanding of geometric figures that goes beyond the formal

---

16. James T. Fey (ed.), *Computing and Mathematics: The Impact on Secondary School Curricula* (Reston, Virginia: National Council of Teachers of Mathematics, 1984), p. 25.

development of Euclidean geometry. They should study the properties of geometric figures under various transformations as well as their measure properties.

Geometric ideas can and should be embedded in mathematics courses at all levels. When studying the order of rational numbers, for example, students can use the geometry of the line to support their understanding of the order properties. In trigonometry, students will understand better such formulas as the formula for the cosine of the difference of two angles if they can relate the formula to a geometric interpretation.

Recent instruction in geometry has tended to place too little emphasis on developing ideas of three-dimensional geometry and promoting students' abilities to represent three-dimensional objects. Students need well-developed spatial abilities for the study of such subjects as engineering, architecture, graphic design, chemistry, biology, physics, geography, meteorology, astronomy, and medicine. The availability of computers with the capability of representing geometric figures may make it easier for teachers to incorporate three-dimensional geometry into the curriculum.

College entrants will need:

- Knowledge of two-dimensional and three-dimensional figures and their properties.
- The ability to think of two-dimensional and three-dimensional figures in terms of symmetry, congruence, and similarity.
- The ability to use the Pythagorean theorem and special right-triangle relationships.
- The ability to draw geometrical figures and use geometrical modes of thinking in solving problems.

Those students who plan to enter fields that will require advanced mathematics courses in college will need additional work in geometry beyond the topics listed above. They will need at least the following:

- Appreciation of the role of proofs and axiomatic structure in mathematics and the ability to write proofs.
- Knowledge of analytic geometry in the plane.
- Knowledge of the conic sections.
- Familiarity with vectors and with the use of polar coordinates.

Functions

The concept of *function* is central to mathematics, and students entering college not only need to understand what functions are in general but also to be familiar with examples such as $f(x) = 3x + 1$ and $g(x) = 5x^2$. They need to be able to evaluate a function when they are given a value from its domain. They also need to be familiar with relations from elementary mathematics, such as *is perpendicular to* and *is a multiple of*. They should be able to use the computer and other tools to represent functions graphically. It is more important for students to have a strong intuitive understanding of various functions and their graphs than it is for them to have learned a formal definition of *function*.

College entrants will need:

- Knowledge of relations, functions, and inverses.
- The ability to graph linear and quadratic functions and use them in the interpretation and solution of problems.

Students who will study advanced mathematics in college need to know about other functions, such as the cube root function, the sine function, and the exponential function. They should learn to represent these functions graphically, using the computer to study the behavior of a function as it is transformed in various ways. They need to know about the composition of functions and the algebra of functions; for example, $f(g(x))$ and $\sin x + \cos x$. These students need the following:

- Knowledge of various types of functions including polynomial, exponential, logarithmic, and circular functions.
- The ability to graph such functions and to use them in the solution of problems.

Planning to Achieve the Outcomes

The two levels of outcomes listed above constitute basic preparation in mathematics for graduates of high school. The outcomes are not radically different from what many secondary schools now offer or could offer easily.

To achieve the outcomes, the syllabi for most high school mathematics courses should be restructured, with some topics eliminated, some deemphasized, and others added. Some proposals for course structure are presented in the next chapter. Many of the preparatory mathematics activities can begin in the seventh and eighth grades. In addition, provisions may need to be made for concurrent course offerings in high school so that students planning to take advanced mathematics in college can be given adequate preparation.

Since the typical college-bound student has not selected a major field of study, all students interested in postsecondary education should be encouraged to continue their study of mathematics through the last year of high school. College-bound students who study no mathematics in the last year or two of high school are likely to have difficulty with the mathematics sections of college entrance examinations. Moreover, some students who do not continue their study of mathematics through the last year of high school may find it necessary to take a remedial mathematics course when they get to college. Even if students are not planning to enter fields that require advanced preparation in mathematics, courses should be available for them to take as seniors so that their mathematical skills and knowledge do not atrophy. If they are not able to fit a mathematics course into their twelfth-grade program, they at least should take a course such as physics or computer science in which mathematics is applied. The high school mathematics program should be organized so that students who make a late decision to pursue postsecondary education are not trapped in dead-end mathematics courses but are able to move into a college preparatory sequence.

The interests of equity require that mathematics teachers make special efforts to encourage students who might be victims of stereotyping regarding their ability to pursue the study of mathematics. Women and members of minority groups, such as blacks and Hispanics, are underrepresented in professions that rely heavily on the mathematical sciences, and they tend to be underrepresented in advanced mathematics courses. Although overt stereotyping of a group as lacking in mathematical ability is no longer common, subtle pressures to discontinue the study of mathematics still exist and can be powerful. Mathematics teachers should be alert to such pressures and should make every attempt to nurture the talents of all students. Mathematics, which so often has been a barrier to college entrance, should become an avenue for providing opportunities to any student who wishes to receive the full benefits of higher education.

Preparation for High School Mathematics

If students are to enter college having achieved the outcomes outlined above, their high school work in mathematics will have to depend on and build on substantial prior preparation. Most of this preparation will be in arithmetic, but some attention also should be given to laying foundations for the study of computing, statistics, algebra, geometry, and functions. The report of the Conference Board of the Mathematical Sciences on the mathematical sciences curriculum for kindergarten to grade 12 contains a useful section on elementary and middle school mathematics that includes many of the recommendations made here.[17]

Students who have not received adequate preparation for high school mathematics will be handicapped in their efforts to achieve the outcomes outlined in the previous sections of this chapter. High school mathematics programs may need to be adapted to the special needs of those students who are poorly prepared. In subsequent chapters we present suggestions for helping such students. As far as possible, however, high school mathematics teachers should try to work together with elementary and middle school teachers to strengthen the mathematics program in the earlier grades and thereby reduce the number of students who enter high school with inadequate preparation in mathematics. Some school districts may find the recommended preparation outlined below difficult to implement, but it should provide a goal toward which the mathematics teachers in a district can work together.

---

17. Conference Board of the Mathematical Sciences, "The Mathematical Sciences Curriculum K-12: What Is Still Fundamental and What Is Not." In National Science Board on Precollege Education in Mathematics, Science, and Technology, *Educating Americans for the Twenty-First Century: Source Materials* (Washington, D.C.: National Science Foundation, 1983), pp. 1-21.

Arithmetic

On entry into high school, all students should understand and be able to perform the basic arithmetic operations with whole numbers. They also need to understand the meaning of fractions, decimals, and percent and their relationship to one another, as well as to be able to perform computations with them. They need facility in rounding, approximation, and numerical estimation, and they should be able to assess the reasonableness of a numerical answer. They need to be able to translate situations and word problems that involve numbers into mathematical statements. They should be able to use and interpret graphs and tables and to compute perimeters, areas, and volumes of simple geometric figures.

Other Topics

Students entering a high school mathematics course should be familiar with the use of calculators and computers in solving problems in arithmetic. They should have used the computer in exploring mathematical ideas by playing appropriate computer games, and they should be able to use a text editor and a document compiler. They should have been introduced to Logo or Karel programming. The programming instruction need not be extensive, but it should include experiences in solving some problems and writing elementary programs, such as games that introduce concepts of sequential execution, procedural abstraction, control structures, and looping.

Students should have performed informal experiments through which they learned about basic ideas of statistics and probability. They should be able to use computers to generate sample spaces and to conduct experiments that previously required cumbersome and time-consuming accumulation of data and exhaustive computations. They should have learned to analyze and interpret newspaper and magazine articles in which statistics are reported. They also should have explored the nature of evidence with experiences in collecting and displaying evidence they have gathered.

Students should be familiar with the use of algebraic symbolism and techniques as generalizations of the numbers and operations of arithmetic. They should be able to identify the standard two-dimensional and three-dimensional geometric figures and the standard formulas for perimeter, area, and volume. They should be familiar with function concepts such as direct and inverse variation, having had experience with dynamic models of increasing and decreasing phenomena as explored with and represented by the computer.

III. The Curriculum

High school mathematics curricula, no matter how they are organized, should permit the school to identify college-capable students at any time in their secondary school career and to place them in appropriate preparatory courses. No student should be denied preparation in mathematics for college solely because of decisions made in the first two years of high school. In the early secondary school years, it is difficult to identify, from a group of average ability, which students will be capable of entering college and succeeding. Provisions need to be maintained for equitable and flexible access to college preparatory courses, and in particular, opportunities should be available for students who may have been shunted into courses such as general mathematics, consumer mathematics, or business mathematics.

The traditional structure of the first three years of college preparatory mathematics in high school has been a year of algebra, a year of geometry, and a second year of algebra. Sometimes the order of the second two years has been interchanged, but this basic set of courses has endured. The arrangement is not especially logical. It arose because colleges first required a year of algebra for admission and later a year of geometry. When they began to require more preparation, the courses moved down in the grades, maintaining their form like successive layers of rock laid down over geologic time. Even the upheaval of the so-called new math did little to affect the course structure of the high school mathematics curriculum in the United States.

As the traditional course structure developed, the first year of algebra became the entry point for college preparation, and alternative courses were created for those students who were deemed unprepared for, or incapable of, the study of algebra. "General mathematics" originally was conceived during the 1920s as a curriculum for grades 7-12 that would integrate algebra, geometry, trigonometry, and elementary statistics, stressing the function concept as a means of unifying these topics. The general mathematics curriculum was widely accepted at grades 7 and 8 but did not dislodge the traditional structure of college preparatory courses at the higher grades. Instead, general mathematics became a course (or two) designed as an alternative to the study of algebra. When schools began to require all students to take at least one year of mathematics in grades 9 to 12, general mathematics became the most common alternative to algebra. Other courses, such as consumer mathematics, business arithmetic, and mathematics for various vocational subjects, became options in many schools. None of these alternative courses was designed for college preparation.

In the curriculum proposed below, we have attempted to bridge the gulf between college preparatory courses and dead-end courses such as general mathematics by formulating a new course, Computational Problem Solving. This course would be designed expressly to prepare students who are not ready to begin the regular college preparatory curriculum. At the same time, we have suggested how the regular curriculum might be restructured so that mathematics is not artificially compartmentalized into year-long units entitled algebra and geometry.

In formulating a proposed curriculum, we have had the guiding principle that *all students who are capable of entering college should be provided as much preparation in mathematics as possible before leaving high school.* It follows, therefore, that *in their academic preparation for college, students should have a continuous experience in mathematics.*

We noted in Chapter 2 that mathematics is both a liberal art and a tool subject. As a liberal art, mathematics should be part of every student's education. When mathematics is divided into first-class and second-class versions, those students who receive the second-class version are cheated. All students should receive essentially the same basic preparation in mathematics, with the principal variation being in the rate at which the mathematics is presented. Students should be provided with a choice from among several mathematics courses only after they have been given a basic preparation. All students deserve an opportunity to become acquainted with the ideas of mathematics.

As a tool subject, the effective study and use of mathematics require that students maintain a high level of skill in the subject. Students who do not study mathematics each year find that their skills deteriorate rapidly. Since students cannot predict with any accuracy the level of mathematical skill that will be required of them in later life, they need to be helped to keep their options open by providing them with as high a level of preparation in mathematics as possible. The continuous study of mathematics in high school can help more students to enter college with the skills needed to begin the study of college mathematics.

We propose a structured set of courses for college preparatory students in which mathematics is taken each year. Depending on the composition of its student body and faculty, and on the needs, wishes, and support of the community, a high school might offer all or part of the proposed curriculum. The curriculum is designed to lead to the outcomes that have been discussed in Chapter 2 while permitting schools considerable flexibility in formulating their college preparatory mathematics program. At the heart of the curriculum is a three-year sequence of courses entitled "Mathematical Topics 1-2-3." This sequence includes the basic topics in computing, statistics, algebra, geometry, and functions that are outlined in Chapter 2. Entry into the sequence is possible at any time during the student's high school career.

Students who are prepared to begin Mathematical Topics 1-2-3 at the end of grade 7 or grade 8 will enter the sequence in either grade 8 or grade 9. If they begin the sequence in grade 8, they will have the opportunity to prepare in grade 11 for taking at least one advanced placement course or other advanced mathematics or computing course in grade 12. Students needing additional preparation or additional academic maturity at the end of grade 8 can take a Computational Problem Solving (CPS) course in grade 9 and enter the sequence in grade 10. Students who begin the sequence later than grade 10 will fall short of meeting all the outcomes specified above. Colleges might require such students to continue their progress before they are admitted or might provide the means for such students to complete their preparation.

Computational Problem Solving (CPS)

The CPS course is intended for students who lack either the preparation or the maturity to begin Mathematical Topics 1-2-3. Such students usually are identified because they are deficient in skills of computational arithmetic; however, greater deficiencies are created by an insufficient understanding of mathematical concepts and processes. The purpose of the course is to improve the students' skill and understanding so as to build a foundation for the study of algebra, geometry, functions, statistics, and computing. The course should expose students to various facets of mathematics and give them some opportunities to experience success with mathematics.

The CPS course would be designed to provide interesting and challenging applications of arithmetic, statistics, and computing to real-world problems. Students would use calculators and, where possible, computers to solve problems. The course should not subject students to lengthy and repeated paper-and-pencil computations. Emphasis would be placed on developing an awareness of the scope and nature of mathematics. The course would be organized with approximately six modules that would be chosen in light of the interests and abilities of the students and teacher.

It is important to stress that the proposed CPS course is not meant to be simply a retreaded course in general mathematics. The course is intended to prepare students for the study of college preparatory mathematics. It is not an arithmetic review course. It is designed to help students who, for whatever reason, are not prepared for algebra and other topics. The course should be forward looking: that is, it should attempt to anticipate key concepts, such as function and variable, and to provide students with a background to understand such concepts through methods largely numerical and computational. The level of abstraction purposely should be kept low, and most of the problem material should be drawn from real situations of interest to students.

At Ohio State University, Joan Leitzel and Alan Osborne are conducting a project to construct four units for grade 7 and four for grade 8 that exemplify the modules one might find in a CPS course. Leitzel and Osborne have taken a numerical problem-solving approach in which calculators and extensive graphical activities are used to provide students with experiences concerning the concept of *variable*. Familiar problem-solving settings are employed, and stress is placed on exploring the idea of functional dependence. The titles and sequence of units are the same for both grades, but the units for grade 8 contain more complex problems. The units are:

1. Using a Calculator. Students learn to use a calculator to explore arithmetic operations and number properties.
2. Solving Problems Numerically. Students investigate problem situations involving numerical relationships by guessing and testing in order to fill out charts.
3. Graphing. Students learn to plot numbers on a number line and to graph relations in the plane.
4. Representing Relationships with Variables. Students use variables to state relationships displayed in charts and graphs constructed earlier.[1]

There is a growing recognition among educators that so-called remedial courses in mathematics often exacerbate the problems they are supposed to solve. Although such courses do have their success stories, too often these courses lower students' expectations, reinforcing their beliefs that they cannot learn mathematics. Some schools are beginning to find that keeping students in regular courses and providing supplementary instruction as it is needed not only may be more cost-effective than providing remedial courses but also may increase achievement.

In line with this reaction against remedial courses, schools might wish to provide additional support to students whose preparation is deficient. Rather than shunting these students into a remedial arithmetic course, schools might wish to keep them in the CPS course or the Mathematical Topics sequence and provide them with special help. That help might consist of a special course run parallel to the regular course, or it might involve special tutoring sessions. The City of Chicago conducts a special assistance program for high school students who are below grade level in mathematics. The regular course meets five days a week and supplemental instruction is given two to five days a week. During the periods of supplemental

---

1. Joan R. Leitzel, draft proposals in personal communication, July 6, 1984.

instruction, high-achieving students are paired with low-achieving students for peer tutoring.[2]

Mathematical Topics 1-2-3

In proposing the set of courses called Mathematical Topics 1-2-3, which ordinarily would span three successive years, we are encouraging schools to organize the topics in computing, statistics, algebra, geometry, and functions given in Chapter 2 in whatever manner they see fit. They can preserve the current algebra/geometry/advanced algebra structure, incorporating computing, statistics, and functions into that structure, or they can construct an integrated mathematics curriculum in which topics in each of the major areas are studied each year and the relationships among topics are stressed.

As the name Mathematical Topics 1-2-3 might suggest, we favor the integration of topics. A more efficient and flexible organization of the content is possible if one is not bound by year-long courses devoted to a single theme. A truly integrated high school curriculum in mathematics would allow computing and statistics to be used as a context for learning algebra; would draw upon geometric concepts to illustrate ideas from algebra, computing, statistics, and functions; and would use functions as a means of unifying much of the other content. Rather than being organized according to concepts and techniques, the curriculum could be organized according to problems, with concepts and techniques from various parts of mathematics introduced as they contribute to understanding and solving a problem. An integrated curriculum can more easily incorporate applications of mathematics and mathematical modeling than can a curriculum divided by content.

We recognize, however, that strong forces of tradition, textbooks, tests, and college requirements—not to mention the powerful force

---

2. Board of Education, City of Chicago, *High School Renaissance Workshop Sessions for High School Principals* (Unpublished monograph, May 1984).

of inertia—may hamper schools in the task of reorganizing the mathematics curriculum for the college-bound. That such a reorganization is not impossible can be seen in the recent experience of New York State in devising a unified mathematics curriculum for its schools. The *Unified Mathematics* books by Howard Fehr and his colleagues[3,4] and the materials produced by the School Mathematics Project in Great Britain[5] also illustrate an integrated approach. Nonetheless, a reorganized curriculum is not attained easily.

Even within the algebra/geometry/advanced algebra structure, much can be done to emphasize the relations between the various parts of mathematics. Some algebra textbooks already contain useful material on computing and statistics. In the geometry course, more emphasis can be placed on topics in analytic geometry, with students using the computer to graph elementary functions and fit lines and simple curves to arrays of statistical data. Although courses might remain organized according to mathematical topic, applications that draw upon different parts of mathematics can be used to help students review various concepts and techniques they have learned in previous courses.

Upper-Level Courses

Many students will take the Mathematical Topics 1-2-3 sequence in grades 9 to 11, leaving them a year for additional preparation in mathematics. Students not planning to pursue careers requiring advanced mathematics might profit from the additional study of topics such as logic, combinatorics, and probability and statistics. They also could study programming methods.

---

3. Howard F. Fehr, James T. Fey, and Thomas J. Hill, *Unified Mathematics: Courses I-III* (Reading, Massachusetts: Addison-Wesley, 1972).

4. Howard F. Fehr, James T. Fey, Thomas J. Hill, and John S. Camp, *Unified Mathematics: Course IV* (Reading, Massachusetts: Addison-Wesley, 1974).

5. Bryan Thwaites, *The School Mathematics Project: The First Ten Years* (Cambridge: Cambridge University Press, 1972).

Students planning to study fields that require advanced preparation in mathematics or computer science might want to take one or more upper-level courses in grade 12. The content of these courses should encompass the topics given under "Additional Preparation" in Chapter 2, including trigonometry, elementary functions, statistics and probability, discrete methods, and programming methods. The courses should be designed to prepare students both for the study of calculus and for the study of computer science. Students who had begun the Mathematical Topics 1-2-3 sequence in grade 8 could, with an appropriate selection of courses in grade 11, take an advanced placement course in calculus or computer science in grade 12.

As high school mathematics faculties devise a selection of mathematics courses for grade 12, they need to remind themselves that among those students in the middle half of the distribution of mathematics achievement in their school are many students with the potential to pursue further work in mathematics in college. The high school should attempt to identify those students with mathematical talent right up to the day of graduation. If the high school does not provide appropriate and challenging courses for these students, many of them will arrive at college unprepared in mathematics and unaware of its role in providing access to various career opportunities.

IV. Teaching Mathematics

This chapter provides some vignettes of teachers who are attempting to revitalize both the curriculum and their instruction by implementing in their classrooms the ideas outlined in Chapters 2 and 3. Each vignette illustrates a strategy for achieving specific teaching goals in mathematics. The vignettes are designed to help other teachers in the task of making college preparatory mathematics a more vital, stimulating subject—one that will attract and prepare students who might be undecided about, or even unsympathetic toward, collegiate study, as well as students who definitely plan to enter college.

We think that vignettes illustrate our message of how the outcomes might be achieved better than discussion of teaching strategies in general terms. Even though a particular vignette might appear somewhat contrived to the reader, we believe that concrete examples carry more weight with teachers than general pronouncements on how teaching should proceed. Teachers talk to one another about teaching in terms of specific situations. They know that what works well in one situation may not work in another. A specific vignette can strike a spark in the teacher who sees how a similar approach might be used in a specific teaching situation.

Double Whammy!

Students are surrounded by statistics in their daily life, but they are seldom aware of how those statistics are used—and misused. Various misuses of statistics can be illustrated by a teacher or a textbook, but students seem to be more sensitive to those misuses they have investigated for themselves. In this first vignette, a mathematics teacher has capitalized on a student's question about representing data to familiarize students with a common misuse of statistics.

The ninth graders in Miss Olmec's Computational Problem Solving course have been using a sports almanac to obtain data for an activity in graphical representation. Karen, Lee, and Jim have been working together to graph world record times for 100-meter free-style swimming, plotting each record time against the year in which the record was set. The students have made separate graphs for men and women and have been sketching curves to fit each set of points, speculating as to when the two curves might meet.

Jim suggests that they superimpose the two graphs; when they do, they discover that the curves intersect in 1980, which should not have happened. Karen observes that the trouble obviously lies in their use of different scales to represent the time in seconds. She asks Miss Olmec whether or not the scales always have to be the same.

Miss Olmec suggests that the three students explore the effects on the graphs, and on the point of intersection, of fixing the unit of time for female swimmers and letting the unit of time for male swimmers vary. When the students have completed their investigation and have written a brief report on their findings, Miss Olmec asks them to report their work orally to the rest of the class. As a follow-up assignment, she asks the students to look for examples of graphs in which curves measured in different units are superimposed in the same graph.

A day later, Andy brings to class a graph from a weekly news-magazine in which a jagged red line showing the varying exchange rate of the English pound in U.S. dollars (with the scale in red at the left) is plotted on the same grid as a jagged blue line showing the number of American tourists, in thousands, who visited the United Kingdom over the same period of time (with the scale in blue on the right).

The class discusses the graph and what it appears to suggest about the relation between the exchange rate and tourism. The students divide into groups to redraw the graph to explore the results of varying the scales. One group's graph shows a modest decline in the exchange rate coupled with a precipitous increase in the number of tourists. A second group's graph shows a sharp decline in the exchange rate accompanied by a gradual rise in the number of tourists. The graph drawn by a third group portrays a relatively constant exchange rate together with a relatively constant number of tourists.

Miss Olmec points out that such graphs are sometimes called "double whammy" graphs because by appropriate adjustments in the two scales, the curves can be made to rise or fall more sharply, creating the impression of drastic changes occurring simultaneously in the two phenomena being compared. In the weeks following this discussion, she posts on the bulletin board—under the heading "Double Whammy!"—graphs that students bring in from various sources to illustrate this manipulation of statistical data.

Monte Carlo

How can a teacher who has a limited knowledge of computing, together with limited access to computers for instruction, help students learn to investigate problems with the help of a computer? The following vignette shows how one teacher managed to overcome such limitations to acquaint students with a powerful technique for simulating the results of an experiment.

Mr. Gimbrell teaches in a small high school whose only instructional computing equipment consists of three Commodore 64 computers, a monitor, and a thermal printer, purchased in 1983 by the Parent-Teachers Association. Mr. Gimbrell has never taken a course in instructional computing, but he has a rudimentary knowledge of BASIC programming that he has acquired on his own. One Friday, when his Mathematical Topics 1 students enter the classroom, they find one of the computers hooked up to a television monitor at the front of the classroom.

Mr. G: "Today, we're going to look at one way to use the computer's power to help us explore the problem of finding the area enclosed by a curve—that is, finding the area of a parabolic segment. The particular problem we'll look at was investigated and solved by Archimedes in the third century B.C. Selma, do you remember what a segment of a circle is?"

Selma: "It's part of a circle."

Mr. G: "What part? Can you show me an example?"

Selma (illustrating with her fingers): "It's the part between a piece of the circle and a chord—the region enclosed by an arc and a chord."

Mr. G: "Fine. Then it makes sense, doesn't it, that a parabolic segment should be the region enclosed by a parabola and a line cutting across the parabola. Like this." (He sketches a parabolic segment on the chalkboard.) "Let's look at the particular parabola $y = x^2$ and try to find the area of the parabolic segment formed by the parabola and the line $y = 1$. Make a careful sketch of that parabolic segment in your notebook. We're interested in the area between the parabola $y = x^2$ and the line $y = 1$.

After the students have drawn the segment in their notebooks, Mr. Gimbrell encourages them to estimate the area of the region—leading them to see that it must be less than the bounding rectangle, whose area is 2, and more than the inscribed isosceles triangle, whose area is 1. He obtains $\frac{3}{2}$, $\frac{4}{3}$, and $\frac{5}{4}$ as estimates, which he writes on the chalkboard. He then goes to the computer console.

Mr. G: "Now, let's take a look at a program for estimating area that Betsy wrote on her computer at home and brought to school last week. She and I have been testing it before class, and it seems to work fine. Let's begin by simplifying our problem. We'll take a triangle whose area we already know—the triangle bounded by the y axis, the line $y = x$, and the line $y = 1$." (He enters this information into the program, and a display appears showing the first quadrant for x and y less than 1 and with the three lines plotted.) "We know the area of the triangle formed by the lines. Jerry, how is it related to the triangle inscribed inside our parabola?"

Jerry: "By the symmetry, it looks like half the inscribed triangle. Its area must be one half."

Mr. G: "That's right. Now, when we run Betsy's program, it will choose points at random with both coordinates between 0 and 1. That means the points will be in the first quadrant, below the line $y = 1$, and left of the line $x = 1$. They'll be somewhere in the square around our triangle. We'll set the number of trials, and whenever the point is inside the triangle, that counts as a hit. Otherwise, it's a miss. Let's start with 50 trials. Keith, how many hits do you think we'll get?"

Keith: "25."

Mr. G: "Why?"

Keith: "Well, maybe not exactly 25. But the number of points should be about 25 because the triangle's area is half the square's."

Anne: "Yes. There's a 50-50 chance that the point will be inside the triangle, so the number of points should be 24, 25, or 26."

Mr. G: "Let's try it." (He enters "50" into the computer, and on the screen the randomly chosen points appear inside the triangle one at a time, while a count of "hits" and "trials" is displayed at the bottom of the screen. When the number of trials reaches 50, the program halts.) "We got 25. Shall we try again?"

After several more trials, Mr. Gimbrell enters the equation for the parabola into the program, noting that, by symmetry, the area of the figure in the first quadrant is half the area they are seeking. For 100 trials, various students guess 75, 67, and 63 hits. When 100 trials yield 66 hits, a discussion begins as to whether or not the parabolic segment is really two-thirds of the bounding rectangle. Since the class period is about to end, Mr. Gimbrell suggests that Betsy run the program at home several times with 100 trials, keeping track of the number of hits. Her data presented in class the next day trigger two discussions—first, a discussion of Monte Carlo Estimation of Area[1]; and second, a discussion of how Archimedes might have solved the problem.

Pizza to Go

Students often possess an academic knowledge of functions without having much idea of how that knowledge can be applied. In this vignette, a teacher uses a realistic situation to help students develop the ability to use linear and quadratic functions in the interpretation and solution of problems. *The students also get experience in formulating problems.*

Four students are working together on a problem about pizzas. They are filling out a table that shows, for pizzas that are 9, 12, 15, and 18 inches in diameter, the amount of crust on each pizza (measured as the perimeter) and the amount of filling in each pizza (approximated by the area). The question they are investigating is

---

1. Howard Kellogg, "In All Probability, a Microcomputer." In Albert P. Shulte and James R. Smart (eds.), *Teaching Statistics and Probability* (Reston, Virginia: National Council of Teachers of Mathematics, 1981 Yearbook), pp. 225-233.

how these amounts change as the diameter of the pizza increases.

Carlos: "Every time the diameter increases, the crust increases by the same amount."

Jona: "No, it doesn't. The crust increases more. Look, when the diameter changed from 9 to 12, the crust changed from 9π to 12π. That's 3π. About 9½ inches."

Carlos: "That's what I meant. Every time the diameter increases 3 inches, the crust increases about 9½ inches. It does it every time. The amount it increases is the same."

Edna: "Every time the diameter increases, the filling increases, too, but it goes up more. You get more filling."

Marlayne: "Here's what we found. For a diameter of 9 inches, the amount of filling is 20.25π square inches; for 12 inches, 36π square inches; for 15 inches, 56.25π square inches; for 18 inches, 81π square inches. Each time it increases by a different amount—a bigger amount."

Edna: "That means that a 12-inch pizza has 3π more crust and 15.75π more filling than a 9-inch pie. And a 15-inch pizza has 3π more crust—the same as a 12-inch pie but 20.25π more filling than a 12-inch pie. Hey, it pays to buy the biggest one you can."

Carlos: "Only if the price is right. Maybe they charge more than four times as much for an 18-inch pizza as for a 9-inch one."

The students decide to investigate both the actual prices for pizzas and what the prices should be if they are set proportional to the areas of the pizzas. The next day, armed with some data from several local pizza stores, the students find discrepancies between the two sets of prices. After some discussion among themselves and with the teacher, they conclude that other factors, such as costs for labor, equipment, and baking, determine the actual prices.

The students draw graphs to show how the crust, filling, and price depend on the diameter of the pizza. They observe that although the price function is somewhat badly behaved, the crust function is linear, and the filling function is quadratic.

The students have been asked to write a problem based on their investigation for another group of students in the class to solve. They decide that questions about prices would not be fair, so they consider the question of how the crust and filling change as the diameter changes.

Jona: "Let's ask them to figure out how much more filling you'd get if you bought an extra large pizza than if you bought a small pizza."

Marlayne: "No, that's too simple. Let's ask them a more difficult question."

Carlos: "I've got it! Let's ask which has more filling—two 9-inch pizzas or one 18-inch pizza?"

Jona: "That's great. I like that."

Edna: "I've got a better idea. Let's ask two questions. When you compare two 9-inch pizzas with one 18-inch pizza, which pie has more filling, and which has more crust? They'll never guess without working it out."

Showing the Flag

Students may fail to see the need to consider matters of design, structure, and style in learning computer programming. As long as the program runs, students may not care if it uses convoluted logic and is not easy to follow. The teacher in this vignette helped a student having difficulty writing a program to become familiar with the value of good programming practice.

A group of students is sitting at computer terminals. Each student is working on developing a program to solve some realistic problem associated with an activity in or around the high school. Kevin has chosen to write a program that will help the marching band set up a formation for their half-time show at the next football game. He begins by attempting a program that will draw an American flag.

Ms. Manero comes over to Kevin's terminal just as the screen erupts in a seemingly random mixture of stars and stripes.

Kevin: "I can't make this program work."

Ms. Manero: "What are you trying to do?"

Kevin: "I can create most of the stripes, but every time I add some stars, it messes up the picture. When I fix the stars, the stripes overlap."

Ms. Manero takes a look at Kevin's program and asks him to explain it to her. As he tries to explain it, Kevin realizes that he has forgotten what one of the variables represents. He also realizes that he no longer remembers the flow of the logic he originally intended to use in drawing the flag.

Kevin: "I remember now. I put these two lines of code right here because I was getting too many stars in the third row, but then I meant to take out this line down here. I must have forgotten—no, I took it out and then put it back in when the red stripe came out in two rows. I'm just not sure now!"

Ms. Manero: "Let me suggest that you print out your program. Then rewrite your program from beginning to end, first in English and then in the programming language."

Kevin: "You mean, I've just been wasting my time up to now?"

Ms. Manero: "No, not at all. But what you *have* been doing is trying to add pieces of code here and there to repair first one bug and then another, and now you've forgotten your original plan. This time, write your plan in English—or half in English and half in programming language—and then convert it to code. Include some comments in your code that will help you remember what you intend to do. Don't try to see how difficult you can make it for someone else to understand your code. Write the code as if you wanted me to be able to figure it out without your help. It should be easy for you to read the code later and recall what you had in mind when you wrote it. I know you think it takes less time to add a few patches than to rewrite the program completely, but that's not always true. If you understand your program well enough to make several meaningful repairs to the details of your code, one after the other, you probably can rewrite it in less time than it takes to make half that many patches."

Kevin: "Thanks. I'll try writing it out."

Ms. Manero: "One other thing. Our compiler allows you to use rather long names for variables. Instead of just using one letter for a variable, try giving each variable a name that will remind you of what it represents. If you do, you won't have the trouble you had a few minutes ago trying to remember what 'B' stands for."

Kevin: "All right, I'll try that too."

Some time later, Kevin asks Ms. Manero to look at his work again.

Kevin: "I got the new program to run without much trouble, but it's still not working right. At least I can explain the program to you, though!"

Ms. Manero: "Kevin, I don't know what's wrong with the logic of the program yet because we haven't been through it. But just by looking at your code I can see that, with all these comments and the long names for variables, we ought to be able to find the flaw if we want to."

Kevin: "If we want to? Of course, we want to—don't we?"

Ms. Manero: "Maybe. But you are at a point now where I think you can learn to apply a general principle that can help you write better codes, not just here, but also in all the programs you write."

By questioning Kevin about the sequence of tasks his program is trying to accomplish, Ms. Manero shows him that the program, as he has planned it, accomplishes just two kinds of basic tasks: (1) drawing a band of color that is one of three solid colors and that has one of three lengths; (2) adding stars to one kind of band without changing the color of the background.

Ms. Manero: "At this point, I could ask you to write the code for the two basic tasks that you want performed; then write more sets of code that make use of those two subprograms to do larger tasks; then write more code using those sets to do higher tasks; and so on, until finally you write the highest level of code that uses the lower modules appropriately to draw the flag. We call this the bottom-up approach. It starts with the most specific task and proceeds to the most general. It's not a bad approach if you do it systematically and if you can keep the final goal in mind all the time. However, it is generally easier and more productive to use the top-down approach."

Kevin: "I see what a bottom-up approach is. It's what I've been using. But what is a top-down approach?"

Ms. Manero: "Just this. Assume that you can do those two basic tasks. In fact, assume that you can do *any* elementary tasks that you might need in drawing the flag. Don't worry about how you would do them; just think of the task as a whole, and ask yourself how it could be broken into simpler parts. For example, how could you break the task of drawing the flag into two simpler parts? What would they be?"

Kevin: "I'm not sure. If you had said three parts instead of two, I would have said the red part, the white part, and the blue part."

Ms. Manero: "But you're still thinking bottom up—you're thinking of the details. Forget the details for a moment. Suppose you were looking at the flag from a distance. Maybe it's blurred a little. What parts would you distinguish?"

Kevin: "I guess I'd see the blue and white corner piece as one part and the red and white stripes as the other."

Ms. Manero: "Fine. Now write a two-line program in which you call the subprograms to draw the two parts."

Kevin: "Just two lines? For subprograms I haven't written?"

Ms. Manero: "Yes. That is the essence of top-down design. You may want more lines for comments to document the two calls, but that's up to you. Just make sure that the next person who reads your program can figure out what those two calls to subprograms are for."

Kevin: "That looks easy. Then what? I still have to write the two subprograms."

Ms. Manero: "Yes, but you have reduced the complexity of your task. You can draw the flag if you can draw the two parts, and each part is simpler than the original task."

Kevin: "I can see that. If I start with the field of stars, how would I write a subprogram?"

Ms. Manero: "Break that simpler task into still simpler tasks. You have to create a rectangular blue background filled with white stars. How can you break that into a sequence of still simpler tasks?"

Kevin: "I can think of two ways. I could fill in the whole blue background all at once and then add all the stars at once, or I could do it one band at a time. I already know how to do that—that's in my old program!"

Ms. Manero: "That's fine."

Kevin: "On the other hand, it seems simpler to do the background all at once and then add the stars all at once. Which would be better?"

Ms. Manero: "That's good thinking. Both ways would have their advantages. One might be easier conceptually, but you already know how to do the details of the other one. I'll leave it up to you."

Kevin goes back to work, having begun to understand the principle that program complexity can be reduced by writing programs in relatively independent modules. The modules can be written in either top-down or bottom-up fashion. Kevin has begun to see some of the advantages of a top-down approach. He also has learned that trying to debug a significant program segment by repeatedly changing isolated pieces of code in an ad hoc fashion can change the original flow of the logic in the segment. It is usually better to rewrite the segment completely, as Ms. Manero asked him to do, keeping in mind both the overall logical flow of the segment as originally conceived and also the bugs that crop up in some of the details. Kevin also has begun to appreciate some of the advantages of including comments and using descriptive names for variables in programs.

Check Digits

Students often get the impression that mathematics exists only in textbooks and consists largely of a set of rules to be memorized. To help overcome that impression, some teachers give assignments in which students investigate on their own various ways in which mathematics is used. In the following vignette, a mathematics teacher explains how he helps students acquire the skills of using resources external to the classroom, write a report based on information from primary and secondary sources, and develop ideas about a topic for the purpose of speaking to a group.

"The students in my Mathematical Topics 2 class are required to write a research paper during the spring term. When I announce the assignment in January, there are always lots of complaints that this is a math course, not an English course, but I don't back down. Instead, I try to provide them with lots of help in identifying a topic and getting started. I recognize that they may not have done much writing in other mathematics classes, but they've done a fair amount in mine, so they know how I value the ability to express mathematical ideas in coherent English sentences. I think their biggest problem is that their view of mathematics is limited by what they have studied in school. Topics such as non-Euclidean geometry, topology, and game theory mean nothing to them. Sometimes I have to do a lot of encouraging to help them begin to investigate a topic."

Mr. Blakemore teaches mathematics to high school sophomores and juniors, and he gives a research paper assignment in every course he teaches. Over the years he has developed a large file of topics, newspaper clippings, magazine articles, and papers written by previous students that he can use to prompt an investigation. However, he encourages students to use their own resources.

His first suggestion is that students watch how numbers or geometric shapes are used in their daily lives. What questions do they have about how some phenomenon works, and how might mathematics be used to investigate that phenomenon? For example, during the last election year, one student studied how information from exit polls is used to estimate who will win the election and how a decision is made to project a winner. Another student looked at systems of preferential voting used in countries such as Australia and some of the paradoxes those systems can produce. Once students pick a question, it usually can be worked into a reasonable topic.

Mr. Blakemore does not pretend to be informed about every topic a student might choose, but he does have ideas about where students might get help, including people in the business and industrial community, faculty members at a nearby college, reference books in the school library, and indexes of periodicals such as *Science*, *Scientific American*, and the *Mathematics Teacher*.

After several weeks, when all students have appeared to make some progress in identifying a topic, Mr. Blakemore asks them to write one-page essays explaining their topics and indicating what reading they have done, who they have consulted, and how they are keeping records of the information they have gathered.

The research paper is due about a month before the end of the spring term. After the papers have been handed in, time is set aside so that each student has five minutes to present orally the main ideas of his or her paper to the class. The mathematics club at the school selects the most promising papers for entry in the state science fair competition the following year.

Mr. Blakemore relates how one student developed a successful project after making an unpromising start: "Joe was having great difficulty finding a topic for his paper. He couldn't think of any good questions that might have mathematical applications. One day he noticed that a library book had 'ISBN' plus a 10-digit number printed on the back cover. 'There's a number,' he said. 'Tell me how to use it as a topic for a research paper.'

"I told him that it might lead to a good paper. Who could tell? I asked him if he knew what ISBN meant (I certainly didn't) and if every book had such a number. Joe said he was just kidding—he didn't see any point in finding out whether or not other books had ISBN numbers and what they might mean.

"I pressed him to at least do some investigation. He found that although not all books have such numbers on the back, they all seemed to have them on the copyright page—at least books published in approximately the last 20 years did. He figured out himself that each title had its own ISBN and that part of the number refers to the publisher.

"The school librarian told Joe that *ISBN* stands for International Standard Book Number, but she didn't know how the number system was set up or why it was separated into four parts. With her help, Joe wrote to several publishers and libraries and found out that the first part of the number identifies the language in which the book is written, the second part identifies the publisher, the third part is a number that the publisher assigns to the book, and the fourth part is a check digit.

"None of us at the school had any idea of what a check digit was, but it sounded interesting. Joe tried the mathematics department at the local community college but did not receive any help. He found a couple of articles in the *American Mathematical Monthly* that discussed check digits as related to error-correcting codes, but the articles were too technical for him to follow. He became discouraged with the topic but, after a lot of looking, he found one article in *Scientific American* and another article in a British journal for mathematics teachers that explained the function of a check digit and how base-11 arithmetic is used in determining the check digit for the ISBN.

"Joe wrote an excellent paper in which he showed how 'casting out nines' can be used in a simple error-correcting code, and then he went on to show how the base-11 system works, how an ISBN check digit is calculated, and how you can decide whether an ISBN you've been given is correct or not. Another student in the class, Bernard, later wrote a computer program that determines the check digit for a given ISBN, and both students entered the local science fair with a joint project on check digits. They won a district award and entered the state finals, but did not win a prize. I don't know what the judges thought about the project, but I think Joe did well for someone who at first wasn't interested in the topic. I know he was very proud of his work. He said it was the first time he had ever done mathematics such as that on his own, and he has gone on to take more mathematics courses in college."

V. Mathematics and the Basic Academic Competencies

As we noted in Chapter 2, the study of mathematics not only reinforces the other Basic Academic Competencies, but it also relies on them. In this chapter, we shall look at each of the Basic Academic Competencies in the light of its interrelation with mathematics, attempting to specify some ways that mathematics teachers can help their students develop and draw upon the basic intellectual skills that they will need in all of their college subjects. Mathematics is an important subject in its own right, but it gains added importance for students when it is taught using an approach that emphasizes its connection with other fields of endeavor.

Reading

On entry into high school, many students have difficulty reading mathematical textbooks. Their problem is likely to be diagnosed as poorly developed reading skills. When help is provided, it often consists of remedial instruction in basic word recognition and literal comprehension. The underlying theory seems to be that if a certain set of basic reading skills is mastered, comprehension will take care of itself.

This "deficit" theory of reading instruction is coming under question.[1] It labels students as "problem readers" who may simply need experience in integrating and transferring their skills to the more sophisticated reading tasks that middle and high school subjects present. The reading of a mathematical text, for example,

---

1. Joan Nelson-Herber and Harold L. Herber, "A Positive Approach to Assessment and Correction of Reading Difficulties in Middle and Secondary Schools." In James Flood (ed.), *Promoting Reading Comprehension* (Newark, Delaware: International Reading Association, 1984).

requires an integration of skills. Students already may possess many of these skills, but they may have had little instruction or practice in integrating them. In elementary school, students learn to read using textbooks in which the situations are familiar and the vocabulary easily understood, but a high school mathematics textbook presents them with vocabulary in a context where skill in pronouncing an unfamiliar word will not help the student establish its meaning. Meaning depends upon experience, and students who lack experience with the concepts being discussed will not be able to understand the discussion, no matter how well their basic word-decoding skills are developed.

The limitations of the deficit model in reading are similar to the limitations of the deficit model in mathematics: students identified as deficient in basic skills are recycled through remedial instruction designed to teach these skills, when their real need is to develop both skills and understanding in the context of interesting new content. Students are labeled as deficient—which sets up a self-reinforcing set of expectations about their performance—and instruction is aimed in the negative direction of overcoming their weaknesses rather than building on their strengths.

Nelson-Herber and Herber argue that "the majority of students who seem to need corrective reading instruction in middle and secondary schools *don't*."[2] Their argument rests on the conviction:

> that *all* students should have the benefit of reading instruction in every classroom where reading is required, and that reading strategies should be taught simultaneously with the content of the subject being taught. If this were done, very few students would need corrective instruction. Unfortunately, we generally abandon the teaching of reading at the very point where students need to integrate the skills gleaned from basic reading instruction with their knowledge, their experience and their reasoning power to address more complex reading comprehension tasks.[3]

Some students, of course, need corrective reading instruction. But that instruction should be designed to complement the reading instruction that is conducted in the mathematics classroom.

---

2. Ibid., pp. 232-244.

3. Ibid., p. 234.

Mathematics textbooks provide rich material for reading instruction. Much of the exposition is dense with new ideas and terminology. Students can write alternative versions in their own words and, by reading each other's work, clarify difficult points. Some teachers have found that having students write their own versions of textbook word problems—as well as having them make up their own problems (as in the "Pizza to Go" vignette in Chapter 4)—sensitizes them to how problems are phrased and how meaning can be changed by a slight rewording.

Because so many students have difficulty reading mathematical material, some textbook authors have attempted to minimize the extent and reading level of the English prose in the book. The prose becomes a sort of Pidgin English. The hope of these authors seems to be that the textbook thereby is rendered easier to use—easier, that is, for the student who does not want to read but only wants an illustration of how to work the exercises in the book. When working the exercises is the primary goal, a book with a minimum amount of exposition allows a lot of room for exercises. But if understanding the ideas behind the exercises is the goal, there is no way—and no reason—to avoid using clear English sentences that convey the ideas, illustrating them and providing a context for them. Fortunately, more mathematics textbooks are beginning to appear in which the English language is used to examine, illustrate, explain, and discuss rather than merely to point.

Students need to learn that there is more to mathematical prose than the exposition in textbooks. Most school libraries have a collection of books on mathematics that are written for the general reader. For example, the *New Mathematical Library,* published by the Mathematical Association of America, contains more than 30 titles on topics of interest to high school students. The National Council of Teachers of Mathematics publishes a list of suggestions for the high school library that teachers and librarians can use as a source for ordering mathematics books. Through assignments in which students locate additional information on a mathematical topic discussed in class, teachers can help students learn how to read expository prose on a mathematical theme.

Students need opportunities to see that not all prose about mathematics has the same density of ideas. They need to learn that they can skim rapidly over an account of how the solution of a particular cubic equation was disputed by Cardano and his contemporaries but that they must read each step slowly and carefully to follow the derivation of a formula to solve the same equation.

Mathematics is a language—or, better said, mathematics is a collection of languages. Each of these languages is more precise than a natural language such as English. The price of precision, however, is a loss of redundancy. Take a word at random out of an English sentence, and you are still likely to get the intended meaning. Take a symbol at random out of an algebraic equation or a line out of a geometric figure, and the meaning inevitably will have been altered, sometimes to the point of unintelligibility.

Reading a mathematical argument, therefore, requires skill in close reading, in making every symbol count. Mathematics teachers should help students appreciate that there is no redundancy in mathematics by showing them how slight alterations have drastic consequences. Computer programming provides an excellent context for illustrating this point. Students readily come to appreciate that computers ordinarily are very unforgiving when they are addressed incorrectly in one of the common programming languages. A student quickly learns that a misplaced semicolon or a misspelled word halts the computer. Mathematics teachers can point out that the same attention that is given to each symbol in writing a computer program ought to be given to each symbol in reading mathematics.

Writing

The "Check Digits" vignette in Chapter 4 illustrates how teachers can use a major assignment to develop skills in writing about a mathematical topic. Writing a research paper gives students practice in formulating a topic, identifying and using various resources, taking notes, organizing and developing a coherent presentation, quoting and paraphrasing accurately, correcting errors and rewriting, and citing reference sources. An important outcome of such an assignment can be increased self-confidence in handling mathematical ideas.

Kathy Woerner[4] has made some useful suggestions about using a research paper as a project in the mathematics classroom. Several years ago, she began assigning a research paper to her students who were enrolled in second-year algebra or a higher-level mathematics course at Westlake High School in Austin, Texas. Her goal was to teach students to read and write mathematics. Today, the assignment has been adopted by other teachers in the school, and it is an established part of the mathematics curriculum for their college-bound students.

The writing assignment should be spread out over several months' time, with intermediate assignments along the way. Like Mr. Blakemore in the "Check Digits" vignette, teachers can ask for a brief essay to see that students have selected a topic and are progressing with their research. Teachers can meet each student in a brief conference to discuss the main elements, not the fine points of writing style, needed for a successful paper. They also can keep a folder of each student's drafts and revisions, so that progress rather than final performance can be the focus of evaluation. The primary audience for the research paper ought to be the other students in the class. Students should read and react to drafts, taking on some of the burden of evaluation. The final papers should be posted or published in some form that allows circulation to the audience of students and to a wider audience if possible. Throughout the process, the teacher should respond to what the student is attempting to say, not simply to what in fact has been written. The ultimate purpose of the assignment should be to demonstrate to students that they have something to say about a mathematical topic.

Writing in a mathematics class should not be confined to major papers. Through the regular use of writing activities, teachers can demonstrate to students how writing interacts with understanding. We sometimes think we have to understand something before we can write about it, but writing is more often an aid to understanding. We may not understand something very well *until* we have written (or spoken) about it. When students keep a journal in a mathematics

---

4. Kathy Woerner, "The High School Mathematics Research Paper," *Mathematics Teacher* 70 (1977): 448-451.

course, they can see for themselves how their understanding of the major ideas in the course changes over the term. When they document a computer program they have written, they learn to anticipate the questions that a reader might raise about how the program operates. When they write brief papers to report on their investigation of a problem, they have the opportunity to reflect on the processes they have used to solve the problem.

Teachers should expect students to provide more than simply the answers to assigned exercises and problems. Students should be encouraged and assisted to write up their solutions, so that someone else can look at a solution and follow the student's line of reasoning. Many teachers have students keep a notebook containing their written assignments rather than having them hand in each day's work. The notebook can be checked at convenient intervals, and it can be used to demonstrate to students the value of recording their work in a form that later will be intelligible to them.

Students need to learn to write for themselves and for their peers. They need opportunities to engage in "prewriting" discussions that will help them shape the message they want to convey and then to present their tentative products to other students for advice and helpful criticism. They need not submit to the teacher everything they write, and the teacher need not feel compelled to grade everything they submit. The teacher should be an editor rather than a judge, and should help students to be editors for each other's work.

Speaking and Listening

At the Open University in England, a course entitled "Developing Mathematical Thinking" has been offered for several years.[5] One of the approaches advocated in that course is captured in the phrase "do, talk, and record": "Children *do* all manner of activities, *talk* about them with each other and their teachers, and *record* some of what happens."[6] The do-talk-record approach allows teachers to

5. Ann Floyd (ed.), *Developing Mathematical Thinking* (London: Addison-Wesley, 1981).

6. Ibid., p. 3.

develop students' skills in speaking and listening—as well as writing and reasoning—in the context of mathematical investigations.

Bob Vertes reported his experience in using the do-talk-record approach in a mixed comprehensive school. He tried to get the classroom to operate in what he called a "research team" atmosphere. He summarized what he expected of his students.

They must be prepared to:

1. Think for themselves—form their own opinions, follow their own logic.
2. Not to be afraid to communicate their thinking for fear it may be wrong.
3. Accept that wrong answers can also be helpful.
4. Listen to their peers, for comments in their own words (which may well be as useful as mine [the teacher's], and perhaps be . . . clearer for being phrased in less formal language).
5. Question their peers' ideas (or any unsupported statements of mine [the teacher's]), asking for justification, example, or proof.
6. Accept that as well as the "simple" problem . . . there may be side issues, or issues to lead us further, perhaps to generalised results.
7. Accept that some correct facts which seem to relate to the situation are not, in fact, useful; that is, that on occasion there is a surfeit of information from which we have to sieve the essentials to combine together to achieve a desired end result.
8. Accept that some problems are open ended; that is, the methods adopted for solution are what matter and not the answers.
9. Accept that more than one method of solution may exist for a problem, and that often these are equally valid.
10. Come to realize that once they understand something it is much easier to recall it when necessary, especially in comparison to a rote formula. The difficulty here is, once you have reached a convincing result, to write reasons and result in the formal language that is used in textbooks, examination papers, and by "mathematicians," and even to use symbolic language, standard notation, to abbreviate this recording of process and results. [7]

Vertes explains to new students what he expects in class: that, when he is talking, he expects them to listen to him, but that he

---

7. Bob Vertes, "Doing, Talking, and Recording with a Whole Class in a Comprehensive School." In Ann Floyd (ed.), *Developing Mathematical Thinking* (London: Addison-Wesley, 1981), pp. 271-272.

also wants to hear them talking and thinking out loud; that their fellow students also will be listening to them and questioning what they have said; and that they must listen to their friends' comments, evaluate them, and raise questions if they have any doubts. He then presents the students with a problem to investigate. Examples are the handshake problem (If 30 people go to a party and each shakes hands once with everyone else, how many handshakes were there all together?) and the circle parts problem (A circular region is divided into how many parts by segments joining n points on the circle?).

Vertes might work with the whole class, or he might divide the class into small groups to work on a problem. Students are encouraged to talk among themselves about the problem, about any patterns they see, and about possible alternative approaches to a solution. The discussion among the students studying perimeter and area in the "Pizza to Go" vignette in Chapter 4 illustrates the discussion that might occur in a small group. When pupils get stuck, they are encouraged to ask themselves self-directing questions that help them clarify the source of their difficulty.

A Study of Exemplary Mathematics Programs, directed by Mark Driscoll of the Northeast Regional Exchange in Chelmsford, Massachusetts, is a qualitative, exploratory investigation funded by the National Institute of Education to identify and describe factors and conditions associated with excellence in precollege mathematics programs.[8] An anecdote from one of the site visits to Trinity High School, a school identified as having an exemplary program, illustrates how students routinely can be given opportunities to discuss mathematical ideas with their peers. In cooperation with a nearby state university, the school participates in a Development Program designed primarily to help enhance the mathematics achievement of students from minority groups. In the program, tutoring and small-group work are provided to students after school hours.

Mr. Robbins, the Development Program coordinator, gave a problem in his Algebra I class and had the students work on the problem in small

8. Mark Driscoll, *Ten Case Studies from a Study of Exemplary Mathematics Programs* (Chelmsford, Massachusetts: Northeastern Regional Exchange, in press).

groups. ("Don't offer *your* answer until you get three people in your group to agree on the answer.") The ensuing discussion was lively and fruitful. (The teacher to us, later: "I couldn't have done that before the Development Program. It showed me what a useful and powerful tool student peer interactions can be.") Later in the same class: "There is a special relationship between the slopes of parallel lines, and the book doesn't say it explicitly. What is it? Don't shout out your answer until you check it out with someone near you."

Peer interactions can be valuable outside class as well. In college, mathematics often becomes a social activity. Students spontaneously form study groups to work on mathematics assignments together. High school students may need to be shown the benefits of collaborative work, and their teachers may need to encourage them to do their homework in pairs or in small groups.

It is possible that with the decline of the daily recitation—in which students were called to the front of the mathematics classroom to present their solution to an assigned problem—skills of speaking and listening began to decline. Few people today would advocate returning to the formal recitation, but many teachers seem to have gone to the other extreme. Their students are never asked to express ideas orally to classmates and the teacher. In such classes, mathematics has become a paper-and-pencil exercise in which one assignment succeeds another in a seemingly endless stream, and nobody except the teacher ever "talks mathematics." The teacher who gets students talking about mathematics finds that the study of mathematics can help improve speaking and listening skills and that speaking and listening can help students improve their understanding of mathematics.

Reasoning

We tend to think of mathematics as a framework of ideas, a discipline, but mathematics has a process dimension as well. Mathematics as a process both requires and develops skill in various types of reasoning. To construct a mathematical model of a real situation, a student must be able to abstract certain critical dimensions of the situation, separating them from irrelevant surface features. Reasoning in mathematics requires the ability to examine a body of information and sort out what is relevant to a given problem and what is not.

Reasoning also relies on discernment in another way. To reason in mathematics, students need to be able to distinguish fact from opinion, conclusion from hypothesis, what has been proved from what has been conjectured. Students need to understand what can count as evidence in an argument and the role of a counterexample in invalidating a line of reasoning. They need to be able to detect fallacies in reasoning.

The skills of reading, writing, speaking, and listening are interwoven with those of reasoning. When students read, they relate what they already know to what is in the text through the use of reasoning skills. When they write, they shape what they have learned, rethinking their arguments and examining their assumptions. Reasoning skills are developed when students engage in conversations about mathematical topics, challenging one another to support their assertions and looking for implicit assumptions. Students exercise and improve their reasoning skills through mental activity. The do-talk-record approach to mathematics instruction discussed in the preceding section provides the kind of mental activity needed for the development of reasoning skills.

A key aspect of reasoning in mathematics is the formulation of mathematical problems. Few teachers or textbooks give explicit attention to how mathematical problems are constructed. Students apparently are supposed to treat the texts as sacred. Yet when mathematics is put into practice, posing the problem so that it lends itself to a mathematical solution is often much more demanding than obtaining a satisfactory solution.

Stephen Brown and Marion Walter[9] discuss two strategies for helping students learn skills of problem posing. The first strategy involves accepting a given datum, observation, assertion, or other phenomenon and considering alternative questions that might be asked about it. For example, one might begin with the sequence of square numbers—1, 4, 9, 16, 25, . . .—and ask questions about the

---

9. Stephen I. Brown and Marion L. Walter, *The Art of Problem Posing* (Philadelphia: Franklin Institute Press, 1983).

difference between successive terms in the sequence, how the sequence might be represented geometrically, whether or not even and odd terms continue to alternate, which terms have exactly 5 divisors, how frequently the digits from 0 to 9 appear in the sequence, and so on. By asking alternative questions about a phenomenon, students come to see that phenomenon in a different light. They note that they often inadvertently impose on a problem narrow constraints that may hamper their search for a solution. They develop greater autonomy in dealing with problems, learning that they can reshape a problem to make it more or less manageable.

The second strategy for developing skills of problem posing is termed the what-if-not strategy. Again, some phenomenon—for example, the theorem of Pythagoras concerning the sides of a right triangle—serves as the starting point. In this case, however, one begins by listing as many attributes of the phenomenon as one can recall. For example, the Pythagorean theorem deals with a right triangle, it deals with squares constructed on the three sides, the variables in the formula $a^2 + b^2 = c^2$ are related by an equal sign, and so on. Then, taking any attribute, one poses the question "What if not?" For example, what if the Pythagorean theorem did not deal with these squares? What would happen if they were rectangles instead? Or semicircles? Or similar polygons? Each what-if-not question gives rise to problems that can be explored further. Students can easily learn to use the what-if-not strategy to pose a host of interesting problems.

Developing skill in problem solving seems to require that students take an active stance in dealing with problems. So many times, when students encounter a problem they cannot solve, they either give up or wait passively for some bright idea to occur to them. They seldom view the problem as something they can manipulate—represent in a different way, simplify, explore for special cases, generalize, or break into parts. Teachers report that experience in transforming problems and in posing problems of one's own appears to help students improve their problem-solving performance. The students learn that they can control the problem rather than allowing the problem to control them.

Reasoning skills that are developed through experience and instruction in mathematical problem solving include skills of plausible reasoning. For example, students learn that a claim such as the assertion that the midpoints of the sides of a quadrilateral form a parallelogram is made more plausible by observing that it holds in the special cases of the rhombus and the rectangle, but it is not thereby proved. Students can reason by analogy to conclude that an infinite sum will behave somewhat like the finite sums with which they are familiar. Students also need to learn, however, that analogy can prove a false friend.

To improve their reasoning abilities in mathematics, students need opportunities to reflect on their work, and they need guidance from their teacher in how to think about their own thinking. Good students usually develop their own techniques for monitoring and controlling their thinking processes, but most students seem to need some instruction and practice. Students who have trouble monitoring their progress in mathematical problem solving, for example, can be given practice in paraphrasing the problem, summarizing what they have done with the problem thus far, predicting what difficulties the problem is likely to present, and discussing with other students various approaches to solving the problem. They can learn to ask themselves questions during problem solving that will help them keep track of their progress. John Mason and his colleagues have developed some useful techniques that help students become their own questioners and that give them suggestions of what to do when they are stuck. Their book *Thinking Mathematically* contains a wealth of interesting problems and suggestions for the teacher who would like to develop reasoning skills in the mathematics classroom.[10]

Studying

Preparing students for college means, in large part, preparing them to learn how to learn. In order to control their own learning, they need confidence that they can learn, but they also need specific skills that they can deploy in learning. Mathematics is a subject in

---

10. John Mason, Leone Burton, and Kaye Stacey, *Thinking Mathematically* (London: Addison-Wesley, 1982).

which students can develop greater confidence in their own intellectual powers, although too often they do not. It is also a subject in which they can learn study skills, but too often no organized attempt is made to teach those skills.

Some study skills have been discussed in other sections of this chapter. Students need, for instance, to learn how to listen in the mathematics class. They need to develop skills in reading and writing mathematical symbolism and prose about mathematics. Other skills, such as taking notes and preparing for examinations, have not been discussed but will be treated later in this section. Instruction in study skills should be seen as building on strengths that students already possess instead of attempting to remedy their deficiencies. When teachers devise activities to illustrate and provide practice in study skills, they need to consider how every student might be helped to achieve some success in each activity.

One set of skills that most students need help with is the setting of goals and priorities for managing their studies. Teachers can help students begin the process of improving these skills through a discussion of how much time various assignments require. Students can be encouraged to estimate how long an assignment will take and then to examine factors that affected the actual time spent. Students need to be more aware of how they study, what study environment works best for them, and how different methods of studying work for them. Teachers can provide specific exercises that will help promote this awareness. They can ask students to identify what they already know about an assignment before they begin and to remind themselves of what they are trying to learn by doing the assignment. Students can be asked to review their homework before bringing it in, summarizing on paper or in discussion with another student what they learned in the homework assignment. A class discussion based on records students have kept of their study behavior can lead students to recognize what methods work best for them for different assignments and can help them improve their use of those methods.

Because the symbolism of mathematics is so highly compressed, learning it requires concentration and reflection. It cannot be learned in a perfunctory way. The student who has only a vague impression of what he or she has read or heard will not be able to master the subject. Students need to recognize in studying mathematics how essential it is to allow time to concentrate on the task at hand and to reflect on what they have done when they have completed the task. The teacher can attempt to model this concentration and reflection in class when solving a problem with students. The teacher can indicate how important it is to concentrate on what the problem says and what it means. When the problem has been solved, the teacher can help the class review the solution, reflecting aloud on what seemed to work, what did not, how the solution might have proceeded differently, how the answer might be checked in various ways, what other kinds of problems might be solved in a similar way, and so on. This process of looking back at a problem will be more effective if, from time to time, the teacher calls the students' attention to what they are doing and how such reflection could help them when solving other problems.

The great value of reviewing one's work cannot be overstressed. When they have completed an assignment, students naturally are inclined to put it aside, especially when the assignment has involved a repetitious string of exercises. By cutting down on the repetition and encouraging students to review their work, teachers promote a more reflective attitude toward mathematics.

The mathematics class can provide students with opportunities to develop skill in using outside resources through assignments such as that discussed in the "Check Digits" vignette in Chapter 4. Students need to learn that not all mathematics is enshrined in their textbook. Assignments that take them out into the community to meet people who have found mathematics useful in their own lives can give students a greater appreciation for the potential value of mathematics. They should learn to use other people as resources. For example, a teacher might point out to a class that new area codes recently have been assigned by the telephone company to several regions of the country. That could raise several questions. How many different telephone numbers are possible with one area code? How many of those numbers does the telephone company use? A call to the telephone company might initiate an interesting investigation dealing with combinations.

The study of mathematics can be organized to help students develop such study skills as taking notes and preparing for examinations. Teachers sometimes assume that their students already know how to take effective notes. But effective note taking requires that one tailor the process to the objective. In this part of the course, should a student take notes? If the answer is yes, the teacher should be prepared not only to explain why but also to provide assistance in taking appropriate notes. It is not enough to say, "Write this down." Teachers need to help students see the structure of an argument by presenting a brief outline before the argument is given. Then the students can rely upon the outline in taking notes. By providing sample notes on what has just been said in a lecture or discussion, teachers can illustrate how to capture important points in various ways—by outlining, by listing key words, by drawing a network of ideas, and so forth. Assignments can be given in which students take notes on a topic in their textbook. The notes then can be compared and used as a basis for discussion in small groups.

Students also need help in how to use their notes when preparing for tests. The exercise of constructing sample test questions based on one's notes helps the student assess the utility of his or her notes and anticipate what questions the teacher might ask. Exercises in constructing different types of test questions give students insight into how questions can be used to assess different aspects of a mathematical topic.

Computer Competency

Computer competency is listed in Chapter 3 of *Academic Preparation for College* as "an emerging need." Throughout this book, we have tried to indicate how the computer can contribute to the learning of secondary school mathematics. We also have suggested that the computer is changing the subject matter of mathematics by stimulating a new interest in questions of discrete, algorithmic mathematics[11,12] and that it will change the subject matter of school mathematics as well.

11. Stephen B. Maurer, "College Entrance Mathematics in the Year 2000," *Mathematics Teacher* 77 (1984a): 422-428.

12. Stephen B. Maurer, "Two Meanings of Algorithmic Mathematics," *Mathematics Teacher* 77 (1984b): 430-435.

Much of school mathematics consists of algorithms—how to add fractions, factor trinomials, bisect line segments, find greatest common divisors, evaluate determinants. An algorithm is simply a procedure for solving a problem in a finite number of steps. Algorithms are so pervasive in school mathematics that when students encounter procedures that are not algorithmic, such as finding the proof of a theorem in geometry, they often feel cheated and seek some foolproof method.

The history of mathematics education can be seen as a history of changes in the algorithmic procedures taught in school. Among the earliest mathematical documents we have are clay tablets recording the efforts of Babylonian schoolboys struggling to multiply and divide in a system in which 60 rather than 10 was the base. The invention of arabic numerals made arithmetic calculations easier to learn, and the rise of commerce in Europe increased the importance of learning "to reckon." Until the adding machine came along, exercises in adding long columns of multidigit numbers were an important part of the school curriculum.

Today the availability of calculators and computers has rendered obsolete many of the algorithmic procedures taught in school. Slide rules and tables of logarithms are disappearing from classrooms. Such techniques as long division should be disappearing too, but their demise will be slower, in part because many people seem to believe that facility in the technique of long division conveys an understanding of the division operation that cannot be obtained in other ways. One of the most critical issues facing mathematics educators over the next decade will be the question of which paper-and-pencil algorithms students should learn and how much they need to practice them in order to understand what function the algorithm performs. If, as many mathematics educators predict, much of the vast amount of time currently spent in learning and practicing algorithmic procedures can be saved with no loss to students' understanding, that time can be put to use in learning other procedures as well as techniques for applying the students' mathematical knowledge.

The availability of computers in mathematics courses can help students develop competency in using computers while they learn mathematics. Students can use computers for word processing and information retrieval as they write about mathematical topics. They can use computers for self-instruction in mathematics, to practice skills and review concepts and principles. They can gain experience in applying mathematics to situations modeled or simulated by means of computers. They can use computers as tools for solving mathematical problems. By learning to develop their own programs, as in the "Showing the Flag" vignette in Chapter 4, students can learn techniques—such as breaking a problem into subproblems, reducing the problem to a simpler or previous case, and working backwards—that apply equally well to programming problems and mathematical problems. The computer allows students to see the experimental side of mathematics. They can investigate number properties and manipulate sequences of numbers, testing various conjectures about the patterns they observe. They can use the graphical capabilities of computers to explore geometric concepts, and they can use geometric notions in learning about computer graphics. Students can learn to apply important mathematical ideas—such as recursion, approximation, and algorithm—to problems in computing. Although all school subjects can make use of computers, mathematics offers special opportunities for students to learn how computers work and to use computers in solving challenging problems.

Observing

Observing is listed in an appendix to *Academic Preparation for College* as "a competency to consider." Mathematics is a subject in which competence in observing can play as vital a role as competence in reading; in fact, much of the reading of mathematical material makes heavy demands on one's ability to observe accurately. Although observing tends to be thought of as a skill primarily connected with science investigations and with creativity in the arts, it is central to the learning of mathematics. Further, by learning techniques of geometric representation that are used in mathematics, the future architect or engineer develops a higher level of visual thinking.

Students can learn to read the visual representations of mathematical ideas to be found in graphs, maps, charts, and diagrams. They develop skill in reading such figures by learning to construct them. Students also can learn various conventions for laying out visual material, particularly the conventions used in representing three-dimensional objects in two dimensions. An investigation of optical illusions can illustrate how mathematics is used to refine representational techniques, as in perspective drawing. Not only can students learn the mathematics of perspective and isometric drawing, they can practice the techniques such drawing requires.

Students need practice in choosing appropriate units for the variables being represented in an accurate figure. Scale drawings require an understanding of proportion, and teachers can help students develop skills in proportional reasoning through activities in which the students draw figures to different scales. The use of computer graphics permits students to see clearly and quickly the effects of changes of scale. The importance of an appropriate and consistent choice of scale was illustrated in the "Double Whammy!" vignette in Chapter 4.

Many students have trouble in geometry when the teacher puts a drawing on the board because they do not understand the role it plays in the problem. They need help in learning to "see the general in the particular." In some problems, the drawing represents "wishful thinking"—if we had what we were looking for, the figure might look like this. Teachers can discuss the potential usefulness of representing both what we know for sure and what we do not know in a particular problem, as well as the importance of keeping these distinct.

Observing in mathematics goes beyond observing geometric figures. By observing patterns of numbers students can discover, and express algebraically, general laws that govern the patterns. Many applications of mathematics require that one find an appropriate function to characterize a set of number pairs. The classroom game of "guess my function," in which students use a table of number pairs to guess the function a classmate has chosen, is essentially an exercise in observing. Students also can learn to observe carefully the data they have gathered, so that they can identify anomalies and select appropriate schemes for summarizing the data. In using calculators and computers, students can learn to be sensitive to various sorts of errors and how those errors are revealed in the output.

The mathematically ignorant person is likely to look at a situation ripe with mathematical meaning in the way a camera might capture it—taking in everything, interpreting nothing. The person educated in mathematics looks at that same situation with the eye of a critic—selecting the essential features and stripping away the irrelevant. Students are not likely to develop the ability to see the world through mathematical lenses unless teachers provide opportunities and direction in applying mathematics to the solution of real problems.

VI. Toward Further Discussion

We recognize that most high school mathematics teachers are struggling to do their best under difficult, and sometimes nearly intolerable, circumstances. We also recognize, however, that there is a cadre of high school mathematics teachers who realize that the college preparatory curriculum needs to be revamped—along with the pedagogy supporting it—and who are willing to invest the time and energy necessary to accomplish this task, with some help and support. These teachers know that the curriculum has become too crowded, that they are able to do justice to only a few of the topics they are supposed to teach, and that after a year or so, if not sooner, their students are able to recall only a fraction of the mathematical concepts and principles they were taught. These teachers also realize that many of their students may have lost interest in mathematics because of the routine, repetitive way in which it is so frequently taught. The results of a series of case studies of secondary schools done in 1976 indicated that students tend to consider their mathematics classes as dull and boring in comparison to their other classes.[1] Even though they readily acknowledge that mathematics is an important subject for them to learn, they do not seem to look forward to studying it.

We would like to invite those teachers who are interested in revitalizing the college preparatory mathematics program to join us in a continuing dialogue on the important issues that must be faced if a curriculum of the sort we have sketched in these pages is to become a reality. In this chapter, we identify and discuss some of the issues we see as important. We recognize, however, that the teachers who respond to our call for a dialogue will have additional issues that they want to place on the agenda. The Educational EQuality Project is a 10-year effort, and this book represents only

---

1. James T. Fey, "Mathematics Teaching Today: Perspectives from Three National Surveys," *Mathematics Teacher* 72 (1979): 490-504.

one marker along the way. We look forward to hearing from interested teachers not only on the issues we discuss here but on other issues that deserve to be considered.

Unresolved Curriculum Issues

In Chapter 3 we attempted to sketch an outline of a college preparatory curriculum that will meet the needs of college-bound students over the next decade. The outline will be elaborated as teachers gain experience in trying out some of our proposals. The following basic assumptions have guided us:

1. Many students who are in the middle half of the distribution of mathematics achievement at the beginning of high school and who are capable of going to college are not well served by current programs.
2. All students who are capable of entering college need a continuous experience in mathematics during high school.
3. The current college preparatory curriculum is too crowded and too fragmented.

It may appear that we simply have made the curriculum more crowded by suggesting more topics. That is not our intent. We contend that, by judicious use of computing technology, much of the time spent on the development of skill in numerical and algebraic manipulation can be used to concentrate on basic ideas and realistic uses of mathematics. The curriculum can be rebuilt around the themes of computing and the solution of applied problems. Content areas such as algebra and geometry that traditionally have been kept separate can be taught together with greater efficiency and mutual reinforcement. The curriculum can be streamlined so that fewer different themes are treated, but they are treated in greater depth.

To help identify and develop the skills of students who arrive in high school without adequate preparation in mathematics, we have suggested the creation of a Computational Problem Solving (CPS) course. However, many students can remain in courses with their peers if they are given additional help. Some communities have arranged for special support in the form of tutoring by members of the school's national honor society or by students from a local college that has "adopted" the school. Some school districts have set up a "hotline" telephone number that students can call for help with their homework. Some schools have set up parallel courses that provide students in need of special help with additional instruction. There are many ways to help the underprepared student who wants to learn mathematics.

The fact is, however, that some students are so inadequately prepared for high school mathematics that teachers and administrators consider an additional year of preparation to be preferable to advancing these students along with their peers. When that decision is made, we want the course for these students to be directed at preparing them for the college preparatory sequence. We do not want the CPS course to become a dead end, a dumping ground. We believe that a computation-based course that relies heavily on a numerical approach to algebraic ideas can provide the needed preparation. We welcome experimentation and dialogue on this point.

The role of geometry in the college preparatory curriculum is a much-debated issue. Although the computer is no panacea, it does offer some prospect of changing the terms of the debate about high school geometry. The computer can allow students to develop their geometric intuition, in two- and three-dimensional settings, and to explore, on their own, many of the properties and theorems traditionally treated in geometry courses. We have set forth a basic set of competencies in geometry that seem reasonable and attainable. We recognize, however, that the question of the relative emphases to be given to synthetic, transformation, coordinate, vector, and differential approaches to geometry is not close to being settled. Consensus probably will be a long time emerging.

The appropriate role of algorithmic, discrete mathematics in the college preparatory curriculum is just beginning to be discussed. It promises to be almost as thorny an issue as that of geometry. Until colleges reach some consensus on the discrete mathematics they expect entering freshmen to know, high schools will find it difficult to decide how much attention to give to such topics as logic, counting, graph theory, and linear programming. In the preceding pages, we have offered some thoughts on discrete mathematics, but much more discussion and some experimentation are needed. More work is needed too on how the various topics that have been suggested for upper-level courses should be treated.

Assessment of Learning Outcomes

If the college preparatory curriculum in mathematics is to change, then assessment practices must change with it. At present, standardized achievement tests are widely used to assess the attainment of curricular goals, but teacher-made quizzes, tests, and examinations also are widely used.

We encourage test developers to respond quickly to progress in the curriculum by incorporating underrepresented topics such as probability and statistics, discrete mathematics, and computing into their tests. For our part, we are looking at ways that College Board examinations designed to assess college preparation in mathematics can be revised to assess better the outcomes we have set forth in this book.

We also are exploring ways in which the assessment of a student's preparation can reflect more of the teacher's professional judgment of the student's achievement. Many of the outcomes we have identified that concern the abilities involved in solving complex problems, applying mathematics to realistic situations, and reporting investigations of mathematical topics are best assessed by teachers. A teacher can make a continuous assessment of a student's work that reflects the student's growth and accomplishments in mathematics over an entire year. A folio of work done in a mathematics course—problems solved, computer programs written, investigations written—accompanied by a teacher's cogent assessment might be helpful to a college mathematics department in deciding where to place a student. Or, a brief form completed by student and teacher to indicate which topics had been studied, together with an assessment of the student's abilities in various domains of mathematics, might accompany an application to college and be considered in the admissions process.

We do not wish to add to teachers' burdens, nor are we under any illusions about the ease of making continuous assessments. But too great an emphasis on multiple-choice and short-answer testing is in part responsible for the fragmented curriculum in school mathematics today. If we are to refocus the curriculum on a smaller number of topics treated in greater depth, the assessment process needs to be refocused as well. For too long, decisions about whether or not students entering college are prepared in mathematics have ignored valuable information from some of the people who are in the best position to provide it—high school mathematics teachers. We invite interested teachers to join us in considering how assessment in college preparatory mathematics might be made more effective.

Teacher Preparation

Until now, we have not discussed the preparation that teachers might need in order to teach the proposed curriculum. We have talked as if the teachers who would like to experiment with our proposals already possessed the requisite knowledge and the experience to proceed.

Some teachers, however, may not be familiar with all the topics we have identified, and others may not be familiar with instructional approaches that integrate the development of academic competencies and the learning of mathematics. These teachers will want some additional preparation. The organization of appropriate and useful in-service activities for teachers is a difficult problem. Over the next few years, we hope to discover and share some models for effective in-service education. We also hope to identify preservice programs that incorporate the approaches and content we have outlined.

Dialogue among Teachers

In the *Study of Exemplary Mathematics Programs*, mentioned in Chapter 5, Mark Driscoll and his colleagues found that in schools identified as having excellent precollege mathematics programs, teachers supported each other by sharing ideas and materials. Collegiality and teamwork seem to be characteristic of the mathematics departments in these schools, usually under the leadership of a talented department head.

Here, for example, are some observations made at Greeley High School, a magnet school for college-bound students, with an enrollment of 1,100, in a large midwestern city:

> At first glance, the stamp of leadership within the mathematics department is not evident. Indeed, after a few interviews the department seemed to run as a cohesive group effort. Eight teachers are involved in mathematics team activities; they discuss courses together, and they even socialize together. And yet, as our visit progressed, it became apparent that the cohesiveness and vibrance we noted in the mathematics department are due in good part to the efforts of Mrs. Wallace, the woman who has been department head since the year after the school's transformation [into a magnet school]. There are some direct measures such as the monthly meetings and her ensuring that honors courses are rotated among staff members, but her leadership can be best characterized by a subtle fostering of trust and mutual respect among the members of the department. . . . We have noticed that teamwork, a partnership of professionals working together, seems to be an essential part of many exemplary mathematics programs. Often hidden in the team process are the quiet efforts of the department head to create and maintain an environment in which teachers are both challenged and able to act in a professional manner. In her conversations with us and in the interactions in which we observed her, Mrs. Wallace showed warm regard and genuine respect for her staff members, and she obviously valued their opinions in both formal and informal departmental meetings. [2]

"A partnership of professionals working together" is essential if any fundamental revision is to occur in a college preparatory program in mathematics. Through cooperative work, teachers can gain the perspective required to see what their program is, what changes need to be made, and how to make them. Outsiders can advise, they can try to help teachers gain that perspective, but they cannot dictate.

When the teachers in a high school mathematics department have begun to work together on the revision of their college preparatory program, they will discover that their colleagues in other

---

2. Mark Driscoll, *Ten Case Studies from a Study of Exemplary Mathematics Programs* (Chelmsford, Massachusetts: Northeastern Regional Exchange, in press).

academic subjects have much to contribute toward helping the study of mathematics both draw upon and enhance the Basic Academic Competencies. The teacher of English can suggest ways of handling a writing assignment so that it emerges from an earlier activity that might justify writing. The teacher of science can suggest how simulation techniques might be used to yield data for a statistical analysis.

High school mathematics teachers also need to work with their colleagues in middle and junior high schools, to discuss the kind of preparation needed for a revised program, and with their colleagues in two-year and four-year colleges and universities, to discuss changes needed in the college mathematics curriculum as the high school mathematics outcomes are achieved.

Indeed, the task of reshaping precollege mathematics has just begun. We of the College Board's Mathematical Sciences Advisory Committee stand ready to work with teachers in these endeavors and will be attempting to carry the work further ourselves. Students deserve a high-quality program that stimulates the development of all their mathematical abilities. They also deserve an equitable program that provides ample opportunity for all to enter college with a strong academic preparation in mathematics. We invite those teachers concerned with improving the preparation in mathematics of every student who aspires to a college education to join us in this task.

Bibliography

Brown, Stephen I., and Marion L. Walter. *The Art of Problem Posing*. Philadelphia: Franklin Institute Press, 1983.

California State Department of Education. *Raising Expectations: Model Graduation Requirements*. Sacramento: California State Department of Education, 1983.

College Board, The. *Academic Preparation for College: What Students Need to Know and Be Able to Do*. New York: College Entrance Examination Board, 1983.

Committee on the Undergraduate Program in Mathematics. *Recommendations for a General Mathematical Sciences Program*. Washington, D.C.: Mathematical Association of America, 1981.

Conference Board of the Mathematical Sciences. "The Mathematical Sciences Curriculum K-12: What Is Still Fundamental and What Is Not." In National Science Board Commission on Precollege Education in Mathematics, Science, and Technology, *Educating Americans for the Twenty-First Century: Source Materials*. Washington, D.C.: National Science Foundation, 1983.

———. *New Goals for Mathematical Sciences Education*. Washington, D.C.: Conference Board of the Mathematical Sciences, 1984.

Daigon, Arthur. "Toward Righting Writing." *Phi Delta Kappan* 64 (1982): 242-246.

Demana, Franklin D., and Joan R. Leitzel. *Transition to College Mathematics*. Reading, Massachusetts: Addison-Wesley, 1984.

Driscoll, Mark. *Ten Case Studies from a Study of Exemplary Mathematics Programs*. Chelmsford, Massachusetts: Northeastern Regional Exchange, in press.

Evans, Christine S. "Writing to Learn in Math." *Language Arts* 61 (1984): 828-835.

Fehr, Howard F., James T. Fey, and Thomas J. Hill. *Unified Mathematics: Courses I-III*. Reading, Massachusetts: Addison-Wesley, 1972.

Fehr, Howard F., James T. Fey, Thomas J. Hill, and John S. Camp. *Unified Mathematics: Course IV*. Reading, Massachusetts: Addison-Wesley, 1974.

Fey, James T. "Mathematics Teaching Today: Perspectives from Three National Surveys." *Mathematics Teacher* 72 (1979): 490-504.

Fey, James T. (ed.). *Computing and Mathematics: The Impact on Second-*

ary School Curricula. Reston, Virginia: National Council of Teachers of Mathematics, 1984.

Floyd, Ann (ed.). *Developing Mathematical Thinking*. London: Addison-Wesley, 1981.

Geeslin, William E. "Using Writing about Mathematics as a Teaching Technique." *Mathematics Teacher* 70 (1977): 112-115.

Joint Committee of the Mathematical Association of America and the National Council of Teachers of Mathematics. *A Sourcebook of Applications of School Mathematics*. Reston, Virginia: National Council of Teachers of Mathematics, 1980.

Kellogg, Howard. "In All Probability, a Microcomputer." In Albert P. Shulte and James R. Smart (eds.), *Teaching Statistics and Probability*. Reston, Virginia: National Council of Teachers of Mathematics, 1981.

King, Barbara. "Using Writing in the Mathematics Class: Theory and Practice." In C. Williams Griffin (ed.), *Teaching Writing in All Disciplines*. San Francisco: Jossey-Bass, 1982.

Margenau, James, and Michael Sentlowitz. *How to Study Mathematics*. Reston, Virginia: National Council of Teachers of Mathematics, 1977.

Mason, John, Leone Burton, and Kaye Stacey. *Thinking Mathematically*. London: Addison-Wesley, 1982.

Maurer, Stephen B. "College Entrance Mathematics in the Year 2000." *Mathematics Teacher* 77 (1984a): 422-428.

———. "Two Meanings of Algorithmic Mathematics." *Mathematics Teacher* 77 (1984b): 430-435.

Nelson-Herber, Joan, and Harold L. Herber. "A Positive Approach to Assessment and Correction of Reading Difficulties in Middle and Secondary Schools." In James Flood (ed.), *Promoting Reading Comprehension*. Newark, Delaware: International Reading Association, 1984.

Pattis, Richard E. *Kârel the Robot*. New York: John Wiley and Sons, 1981.

Ralston, Anthony, and Gail S. Young. *The Future of College Mathematics*. New York: Springer-Verlag, 1983.

Sachs, Leroy (ed.). *High School Student Merit Awards*. Reston, Virginia: National Council of Teachers of Mathematics, 1984a.

———. *Middle School Student Merit Awards*. Reston, Virginia: National Council of Teachers of Mathematics, 1984b.

Sharron, Sidney, and Robert E. Reys (eds.). *Applications in School Mathematics*. Reston, Virginia: National Council of Teachers of Mathematics, 1979.

Shulte, Albert P., and James R. Smart (eds.). *Teaching Statistics and Probability*. Reston, Virginia: National Council of Teachers of Mathematics, 1981.

Tanur, Judith M., Frederick Mosteller, William H. Kruskal, Richard F. Link, Richard S. Pieters, and Gerald R. Rising (eds.). *Statistics: A Guide to the Unknown*. San Francisco: Holden-Day, 1972.

Thwaites, Bryan. *The School Mathematics Project: The First Ten Years*. Cambridge: Cambridge University Press, 1972.

Tobin, Catherine D. *hm Math Study Skills Program* (student text and teacher's guide). Reston, Virginia: National Council of Teachers of Mathematics and National Association of Secondary School Principals, 1980.

Vertes, Bob. "Doing, Talking, and Recording with a Whole Class in a Comprehensive School." In Ann Floyd (ed.), *Developing Mathematical Thinking*. London: Addison-Wesley, 1981.

Watson, Margaret. "Writing Has a Place in a Mathematics Class." *Mathematics Teacher* 73 (1980): 518-519.

Woerner, Kathy. "The High School Mathematics Research Paper." *Mathematics Teacher* 70 (1977): 448-451.

Members of the Council on Academic Affairs, 1983-85

Peter N. Stearns, Heinz Professor of History, Carnegie-Mellon University, Pittsburgh, Pennsylvania (*Chair* 1983-85)

Dorothy S. Strong, Director of Mathematics, Chicago Public Schools, Illinois (*Vice Chair* 1983-85)

Victoria A. Arroyo, College Board Student Representative, Emory University, Atlanta, Georgia (1983-84)

Ida S. Baker, Principal, Cape Coral High School, Florida (1984-85)

Michael Anthony Brown, College Board Student Representative, University of Texas, Austin (1983-85)

Jean-Pierre Cauvin, Associate Professor of French, Department of French and Italian, University of Texas, Austin (1983-84)

Alice C. Cox, Assistant Vice President, Student Academic Services, Office of the President, University of California (1983-84, Trustee Liaison 1984-85)

Charles M. Dorn, Professor of Art and Design, Department of Creative Arts, Purdue University, West Lafayette, Indiana (1983-84)

Sidney H. Estes, Assistant Superintendent, Instructional Planning and Development, Atlanta Public Schools, Georgia (1983-85)

David B. Greene, Chairman, Division of Humanities, Wabash College, Crawfordsville, Indiana (1984-85)

Jan A. Guffin, Chairman, Department of English, North Central High School, Indianapolis, Indiana (1983-85)

John W. Kenelly, Professor of Mathematical Sciences, Clemson University, South Carolina (1983-85)

Mary E. Kesler, Assistant Headmistress, The Hockaday School, Dallas, Texas (Trustee Liaison 1983-85)

Arthur E. Levine, President, Bradford College, Massachusetts (1983-85)

Deirdre A. Ling, Vice Chancellor for University Relations and Development, University of Massachusetts, Amherst (Trustee Liaison 1983-84)

Judith A. Lozano-Loredo, Superintendent, Southside Independent School District, San Antonio, Texas (1983-84)

Eleanor M. McMahon, Commissioner of Higher Education, Rhode Island Office of Higher Education, Providence (1984-85)

Jacqueline Florance Meadows, Instructor of Social Science, North Carolina School of Science and Mathematics, Durham (1983-84)

Michael J. Mendelsohn, Professor of English, University of Tampa, Florida (1983-84)

Fay D. Metcalf, History Coordinator/Teacher, Boulder High School, Colorado (1983-85)

Vivian Rivera, College Board Student Representative, Adlai E. Stevenson High School, Bronx, New York (1984-85)

Raul S. Rodriguez, Chair, Language Department, Xaverian High School, Brooklyn, New York (1984-85)

Michael A. Saltman, Chairman, Science Department, Bronxville School, New York (1983-85)

Vivian H. T. Tom, Social Studies Teacher, Lincoln High School, Yonkers, New York (Trustee Liaison 1983-84)

Kenneth S. Washington, Vice Chancellor for Educational Services, Los Angeles Community College District, California (1983-85)

Henrietta V. Whiteman, Director/Professor, Native American Studies, University of Montana, Missoula (1984-85)

Roberto Zamora, Deputy Executive Director, Region One Education Service Center, Edinburg, Texas (1984-85)

117592

510.71273 K   117592

Kilpatrick, Jeremy.

Academic preparation in
 mathematics

510.71273 K   117592

Kilpatrick, Jeremy.

Academic preparation in
 mathematics

DATE DUE	BORROWER'S NAME